DIFFUSIONS IN ARCHITECTURE

II Æ

Also by Matías del Campo

Evoking through Design - Contemporary Moods in Architecture - AD Wiley 2017

Machine Hallucinations - Architecture and Artificial Intelligence (Co-edited with Neil Leach) - AD Wiley 2022

IV Æ

DIFFUSIONS IN ARCHITECTURE

-

ARTIFICIAL INTELLIGENCE AND IMAGE GENERATORS

FIRST EDITION

Edited By
Matias del Campo ed.

WILEY

Published by John Wiley & Sons, Inc., Hoboken, New Jersey.
Published simultaneously in Canada.

For general information on our other products and services or for technical support, please contact our Customer Care Department within the United States at (800) 762-2974, outside the United States at (317) 572-3993 or fax (317) 572-4002.

Wiley also publishes its books in a variety of electronic formats. Some content that appears in print may not be available in electronic formats. For more information about Wiley products, visit our web site at www.wiley.com.

Library of Congress Cataloging-in-Publication Data Applied for:
Paperback ISBN: 9781394191772

Graphic Design: Matias del Campo
Cover Image: Courtesy of Matias del Campo

This project was made possible with funding from the Taubman College of Architecture and Urban Planning at the University of Michigan.

SKY10064453_011024

To those who teach machines to think and feel,
Who embrace the estrangements of paths yet to be taken,
And show us how imagination can be real.

"Diffusion models offer a powerful new approach to generative modeling, enabling us to capture the complex dynamics of real-world phenomena in a way that was not previously possible. As a result, we can create more nuanced and sophisticated aesthetics in our generative outputs."

Yann LeCun

"The creative imagination is the synthesis of existing elements into a new whole."

Henri Bergson

"A generative model is like having a genie in a bottle which can make all your wishes come true, but you are facing the problem that you still have to find out yourself what it is that you actually want."

Mario Klingemann

"Synthetic imagination is the art of taking two ideas that have never met before and creating a new idea."

Albert Einstein

X Æ

Acknowledgments

The completion of this book is the culmination of a fast paced journey that has been both rewarding and challenging, yet one that I would not have traversed without the invaluable support and guidance of those who have walked alongside me.

To the architects, designers, and thinkers whose creativity and passion have inspired me, I offer my heartfelt thanks. Your work has been a guiding light, illuminating the path forward for those of us who seek to push the boundaries of what is possible in reimagining the built environment through the lens of artificial intelligence. First and foremost, the architects who made this book possible: Cesare Battelli, Kory Bieg, Daniel Bolojan, Niccolo Casa, Virginia San Fratello, Soomeen Hahm, Immanuel Koh, Andrew Kudless, Elena Manferdini, Ryan Vincent Manning, Sandra Manninger, Sina Mostafavi, Rasa Navasaityte, Igor Pantic, Kyle Steinfeld, Marco Vanucci, and Dustin White, I owe you. To the scholars and researchers who have contributed with thoughtful texts unpacking the problem at hand, and whose rigorous inquiries have informed the ideas presented in this book, I am deeply grateful. Your insights have been a source of clarity and perspective, and have helped to shape my own thinking on this complex and ever-evolving field. Thank you Mario Carpo, Bart Lootsma, and Joy Knoblauch for your invaluable contributions. To the AI experts whose technical expertise has enabled the realization of new forms, functions, and aesthetics, I extend my appreciation. Your contributions have been indispensable in translating conceptual ideas into tangible realities. Alexandra Carlson, Danish Syed, Janpreet Singh, Justin Johnson, Jessy Grizzle, and the many others who have shared their knowledge within the AR^{2}IL laboratory I thank you. Without the support that I have experienced through the Taubman College of Architecture and Urban Planning at the University of Michigan, this book would not have been possible. My thanks go out in particular to Dean Jonathan Massey for his continuous support of my work and to Associate Dean of Research Kathy Velikov and her team at R+CP for making this publication and many others of my endeavors possible. To Vishal Rohira, Fatima Azahra Addou, and Siya Sha for helping to make this book a reality. To the readers who engage with this work, I invite you to journey with me through a landscape of synthetic imagination, and to explore the possibilities that arise at the intersection of architecture and AI. Finally, to my loved ones who have supported me my entire life, providing encouragement and understanding in equal measure, I offer my heartfelt gratitude to you, Sandra Manninger, my wife, partner in crime, and inspiration for everything I do. Mother, sister: your unwavering love has been a constant source of motivation and has sustained me through the highs and lows of every creative process.

In the words of Ernest Hemingway, *The world breaks everyone, and afterward, some are strong at the broken places.* This book is the product of the creative and intellectual challenges that have tested me, and the support that has strengthened me. To all those who have helped me on this journey, I extend my deepest thanks.

With gratitude and humility,

Matias del Campo

Contents

Ⓘ

ꟸE

///// Appendix /////

Preface

Lev Manovich

Introduction to Diffusion in Architecture

This book is a unique document of the beginning of a true revolution in cultural imagination and creation. This revolution has been in development for over 20 years. The first AI papers proposing that the web universe of texts, images, and other cultural artifacts can be used to train computers to do various tasks appeared already in 1991–2001. In 2015, Google's *deep dream* and *style transfer* methods attracted lots of attention: suddenly computers could create new artistic images mimicking the styles of many famous artists. The release of DALL-E 2 in 2021 was another milestone: now computers could synthesize images from text descriptions. MidJourney, Stable Diffusion, and DALL-E 2 all contributed to the acceleration of this evolution in 2022. Synthetic images could not have many aesthetics that range from photo realism to any kind of physical or digital medium, including mosaics, oil paintings, street photography, or 3D CG rendering. The code for producing such images (also referred to as a *model* in the field of artificial intelligence) was made public in August 2022, sparking a flurry of experiments and accelerating development.

I've been using computer tools for art and design since 1984, and I've seen a few major media revolutions, including the introduction of Mac computers, the development of photorealistic 3D computer graphics and animation, the rise of the web after 1993, and the rise of social media sites after 2006. The new AI *generative media* revolution appears to be as significant as any of them. Indeed, it is possible that it is as significant as the invention of photography in the 19th century or the adoption of linear perspective in Western art in the 16th century. In what follows, I will describe a few aspects of AI visual generative media (as of Spring 2023) that I believe are particularly significant or novel. But first, let's define our terms.

The Terms

In this introduction, *artist* or *creator* refers to any skilled person who creates cultural objects in any media or their combinations – architects and designers are included. Indeed, because many architects and architecture students are now experimenting with AI image and animation generation, the old concept of *paper architecture* is resurfacing. Let's face it: building architecture is only a small part of what architects imagine, design, and debate. Images that articulate new ideas, aesthetics, ideals, and arguments have always been the primary focus of the architecture field, and we should not be ashamed of this. So, in my opinion, this book contains real architecture because it raises many interesting questions and points of view. It is unimportant whether some of the ideas expressed in these images will be realized as architecture for VR or physical spaces.

The terms *generative media*, *AI media*, *generative AI*, and synthetic media are all interchangeable. They refer to the process of creating new media objects with deep neural networks, such as images, animation, video, text, music, 3D models and scenes, and other types of media. Neural networks are also used to generate specific elements and types of content, such as photorealistic human faces and human poses and movements, in addition to such objects. They can also be used in media editing, such as replacing a portion of an image or video with another content that fits spatially. These networks are trained on vast collections of media objects already in existence. Popular artificial neural network types for media generation include diffusion models, text-to-image models, generative adversarial networks (GAN), and transformers. For the generation of still and moving images using neural networks, the terms *image generation*, *synthetic image*, *AI image*, and *AI visuals* can be used interchangeably.

Note that the word *generative* can also be used in different ways to mean making cultural artifacts using any algorithmic process (not just neural networks) or even a rule-based process that doesn't use computers. This is how the terms *generative art* and *generative design* are often used in popular culture and the media today. In this introduction, I use the word *generative* in a narrower sense to refer to deep network methods to make media artifacts and apps that use these methods.

AI as a Cultural Perception

There is no one specific technology or a single research project called *AI*. It is our cultural perception that evolves over time. When an allegedly uniquely human ability or skill is being automated, we refer to it as *AI*. As soon as this automation is successful, we stop referring to it as an AI case. In other words, AI refers to technologies and methodologies that are starting to function but aren't quite there yet.

AI was present in the earliest computer media tools. The first interactive drawing and design system, Ivan Sutherland's Sketchpad (1961–1962), had a feature that would automatically finish any rectangles or circles you started drawing. In other words, it knew what you were trying to make. So this was undoubtedly *AI* already. My first experience with a desktop paint program running on Apple II was in 1984, and it was truly amazing to move your mouse and see simulated paint brushstrokes appear on the screen. But today, we no longer consider this AI. Or, for example, a Photoshop function that automatically selects an outline of an object that was added many years ago – this, too, is AI. The history of digital media systems and tools is full of such AI moments – amazing at first, then taken for granted and forgotten as *AI* after a while. (In AI history books, this phenomenon is referred to as the *AI effect.*) At the moment, creative AI/artistic AI stands for recently developed methods where computers transform some inputs into new media outputs (e.g., text-to-image models) and specific techniques (e.g., certain types of deep neural networks). However, we must remember that these methods are neither the first nor the last in the long history and future of simulating human art abilities or assisting humans in media creation.

From Representation to Prediction: AI images and Media History

Historically, humans created images of existing or imagined scenes by a number of methods, from manual drawing to 3D CG (see below for an explanation of the methods). With AI generative media, a fundamentally new method emerges. Computers use large datasets of existing representations in various media to predict new images (still and animated). I can certainly propose different historical paths leading to visual generative media today, or divide one historical time line into different stages, but here is one such possible trajectory:

I Creating representations manually (e.g., drawing with various instruments and carving). More mechanical stages and parts were sometimes executed by human assistants typically training in master's studio – so there is already some delegation of functions.

II Creating manually but using assistive devices (e.g., perspective machines and camera lucida). From hands to hands + device. Now some functions are delegated to mechanical and optical devices.

III Photography, X-ray, video, volumetric capture, remote sensing, photogrammetry. From using hands to recording information using machines. From human assistants to machine assistants.

IV 3D CG. You define a 3D model in a computer and use algorithms that simulate effects of light sources, shadows, fog, transparency, translucency, natural textures, depth of field, motion blur, etc. From recording to simulation.

V Generative AI – using media datasets to predict still and moving images. From simulation to prediction.

Prediction is the actual term often used by AI research in their publications describing visual generative media methods. So while it can be used even evocatively, actually this is what happens scientifically when you use image-generative tools. If you are working with text-to-image model, the net attempts to predict the images that correspond best to your text input.

I am certainly not suggesting that using all other already-accepted terms such as *generative media* is bad. But if we want to better understand the difference between AI visual media synthesis methods and other representational methods developed in human history, *prediction* well captures this difference.

Mapping Between Media

There are several methods for creating AI media. One method transforms human media input while retaining the same media type. Text entered by the user, for example, can be summarized, rewritten, expanded, and so on. The output, like the input, is a text. Alternatively, in the image-to-image generation method, one or more input images are used to generate new images.

However, there is another path that is equally intriguing from the historical and theoretical perspectives. AI media can be created by automatically *translating* content between media types. Because this is not a literal one-to-one translation, I put the word *translation* in quotes. Instead, input from one medium instructs a neural network to predict the appropriate output from another. Such input can also be said to be *mapped* to some outputs in other media. Text is mapped into new styles of text, images, animation, video, 3D models, and music. The video is converted into 3D models or animation. Images are *translated* into text and so on. Text-to-image method translation is currently more advanced than others, but they will catch up eventually. Translations (or mappings) between one media and another were done manually throughout human history, often with artistic intent. Novels have been adapted into plays and films, and comic books have been adapted into television series. A fictional or non fictional text is illustrated with images. Each of these translations was a deliberate cultural act requiring professional skills and knowledge of the appropriate media.

Some of these translations can now be performed automatically on a massive scale thanks to AI, becoming a new means of communication and culture creation. What was once a skilled artistic act is now a technological capability available to everyone. We can be sad about everything that will be lost as a result of the automation and democratization of this critical cultural operation – skills, originality, *deep creativity*, and so on. However, any such loss may be only temporary if culture AI development improves its abilities to be original and understand context.

Because the majority of people in our society can read and write in at least one language, text-to-another media methods are currently the most popular. Text-to-image, text-to-animation, text-to-3D model, and text-to-music are among them. These AI tools can be used by anyone who can write, or by using Google Translate to create a prompt in a language these tools understand well, such as English. However, other media mappings can be equally interesting for professional creators. Throughout the course of human cultural history, various translations between media types have attracted attention. They include translations between video and music (club culture); long literary narratives turned into movies and television series; any texts illustrated with images in various media such as engravings; numbers turned into images (digital art); texts describing paintings (ekphrasis, which began in Ancient Greece); and mappings between sounds and colors (especially popular in modernist art).

The continued development of AI models for mappings between all types of media, without privileging text, has the potential to be extremely fruitful, and I hope that more tools can accomplish this. These tools can be used alone or in conjunction with other tools and techniques. I am not claiming that will be able to create innovative interpretations of *Hamlet* by avant-garde theater directors such as Peter Brook or astonishing abstract films by Oscar Fishinger that explored musical and visual correspondences. It is sufficient that new media mapping AI tools stimulate our imagination, provide us with new ideas, and enable us to explore numerous variations of specific designs.

Simulation and Originality

Both the modern human creation process and the AI generative media process seem to function similarly. A neural network is trained using unstructured collections of cultural content, such as billions of images and their descriptions or trillions of web and book pages. The net learns associations between these artifacts' constituent parts (such as which words frequently appear next to one another) as well as their common patterns and structures. The trained net then uses these structures, patterns, and *culture atoms* to create new artifacts when we ask it to. Depending on what we ask for, these AI artifacts might closely resemble what already exists or they might not.

Similarly, our life is an ongoing process of both supervised and unsupervised cultural training. We take art and art history courses; view websites, videos, magazines, and exhibition catalogs; visit museums; and travel in order to absorb new cultural information. And when we *prompt* ourselves to make some new cultural artifacts, our own nervous networks (infinitely more complex than any AI nets to date) generate such artifacts based on what we've learned so far: general patterns we've observed, templates for making particular things (such as drawing a human head with correct proportions or editing an interview video), and often concrete parts of existing artifacts. In other words, our creations may contain both exact replicas of previously observed artifacts and new things that we represent using templates we have learned, such as color combinations and linear perspective. Additionally, both human and AI models frequently have a default *house* style (the actual term used by Midjourney developers). If you didn't specify the subject yourself, AI will generate it using this aesthetic. A description of the medium, the kind of lighting, the colors and shading, and/or a phrase like *in the style of* followed by the name of a well-known artist, illustrator, photographer, fashion designer, or architect are examples of such specifications.

Because it can simulate tens of thousands of already-existing aesthetics and styles and interpolate between them to create new hybrids, AI is more capable than any single human creator in this regard. However, at present, skilled and highly experienced human creators also have a significant advantage. Both humans and artificial intelligence are capable of imagining and representing both nonexistent and existing objects and scenes. However, human-made images can include particular content, certain details, and distinctive aesthetics that are currently beyond the capabilities of AI. In other words, a large group of highly skilled and experienced illustrators, photographers, and designers can represent everything a trained neural net can do (although it will take much longer), but they can also visualize objects and compositions and use aesthetics that the neural net cannot do at this time.

What is the cause of the aesthetic and content gap between human and artificial creators? Most frequently occurring cultural *atoms*, structures, and patterns in the training data are successfully learned during the process of training an artificial neural network. In the *mind* of a neural net, they gain more importance. On the other hand, atoms and structures that happen very infrequently or only once are not learned. They do not enter the artificial culture universe as learned by AI. Consequently, when we ask AI to synthesize them, it is unable to do so.

Because of this, Midjourney, Stable Diffusion, or RunwayML are not currently able to generate drawings in my style, expand my drawings by adding newly generated parts, or replace specific portions of my drawings with new content drawn in my style (e.g., perform *outpainting* or *inpainting*). Instead, AI generates more generic, common objects than what I frequently draw when I attempt to do such operations. Or it produces something that is merely ambiguous but uninteresting.

I am certainly not claiming that the style and the world shown in my drawings is completely unique. They are also a result of specific cultural encounters I had, things I observed, and things I noticed. But because they are uncommon (and thus unpredictable), AI finds it difficult to simulate them, at least without additional training using my data. Here, we encounter the greatest obstacle we face as creators in using the AI-generated media. Frequently, AI generates new media artifacts that are more generic and stereotypical than what we intended. This may include elements of content, lighting, crosshatching, atmosphere, spatial structure, and details of 3D shapes, among others. Occasionally, it is immediately apparent, in which case you can either attempt to correct it or disregard the results. Very often, however, such *substitutions* are so subtle that we cannot detect them without extensive observation or, in some cases, the use of a computer to analyze quantitatively numerous images.

In other words, new AI generative media models, like the discipline of statistics since its inception in the 18th century and the field of data science since the end of the 2010s, deal well with frequently occurring items and patterns in the data, but do not know what to do with the infrequent and uncommon. We can hope that AI researchers will be able to solve this problem in the future, but it is so fundamental that we should not anticipate a solution immediately.

Content vs Style in AI Images

In the arts, the relationship between *content* and *form* has been extensively discussed and theorized. This brief section does not attempt to engage in all of these debates or to initiate discussions with all relevant theories. Instead, I'd like to consider how these concepts play out in AI's *generative culture*. But instead of using content and form, I'll use different pairs of terms that are more common in AI research publications and online conversations between users. They are *subject* and *style*.

At first glance, AI media tools appear capable of clearly distinguishing between the subject and style of a representation. In text-to-image models, for instance, you can generate countless images of the same subject. Adding the names of specific artists, media, materials, and art historical periods is all that is required for the same subject to be represented differently to match these references.

Photoshop filters began to differentiate between subject and style in the 1990s, but AI-generative media tools are more capable. For instance, if you specify *oil painting* in your prompt, simulated brushstrokes will vary in size and direction across a generated image based on the objects depicted. AI media tools appear to *understand* the semantics of the representation, as opposed to earlier filters that simply applied the same transformation to each image region regardless of its content. For instance, when I used a painting by Malevich and a painting by Bosch in a prompt, AI generated an image of space that contained Malevich-like abstract shapes as well as many small human and animal figures that were properly scaled for perspective.

AI tools routinely add content to an image that I did not specify in my text prompt, in addition to representing what I requested. This frequently occurs when the prompt includes *in the style of* or *by* followed by the name of a renowned visual artist or photographer. In one experiment, I used the same prompt in Midjourney AI image tool 148 times, each time adding the name of a different photographer. The subject in the prompt renamed always the same – empty landscape with some builds, a road, and electric poles with wires going to horizon. Sometimes adding a photographer's name had no effect on the elements of a generated image that fit our concept of style, such as contrast, perspective, and atmosphere. Every now and again, Midjourney also modified image content. For example, when well-known photographs by a particular photographer feature human figures in specific poses, the tool would occasionally add such figures to my photographs. (Like Malevich and Bosch, they were transformed to fit the spatial composition of the landscape rather than mechanically duplicated.) Midjourney has also sometimes changed the content of my image to correspond to a historical period when a well-known photographer created his most well-known photographs. Here's another thing I noticed. When we ask Midjourney or a similar tool to create an image in the style of a specific artist, and the subject we describe in the prompt is related to the artist's subjects, the results can be very successful. However, when the subject of our prompt and the imagery of this artist are very different, *rendering* this subject in this style frequently fails.

To summarize, in order to successfully simulate a given visual style using current AI tools, you may need to change the content you intended to represent. Not every subject can be rendered successfully in any style. This observation, I believe, complicates the binary opposition between the concepts of *content* and *style*. For some artists, AI can extract their style from examples of their work and then apply it to different types of content. But for other artists, their style and content can't be separated. For me, these kinds of observations and subsequent thoughts are one of the most important reasons for using new media technologies like AI-generative media and learning how they work. Of course, I had been thinking about the relationships between subject and style (or content and form) for a long time, but being able to conduct systematic experiments like the one I described brings new ideas and allows us to look back at cultural history in new ways. And this is why I'm particularly excited to write this introduction to the book that brings together the ideas, experiments, and explorations of a number of architects who are part of the current AI revolution.

Prologue: The Weird Ontology of Diffusion Models

Matías del Campo

The idea for this book started to take shape in July and August 2022. It was at this time that the phenomenon of employing diffusion models to speculate about architecture began to snowball, growing larger with each passing day. The proliferation of images across social media platforms, coupled with the lightning-fast speed at which dedicated channels were emerging to showcase architectural phantasms created with the aid of text-to-image generators, served to cement the need to create a record of the explosive burst of diffusion models into the architecture scene. The advent of these tools marks a significant turning point in the relationship between human creativity and algorithmic intelligence (AI), opening up a new realm of architectural possibility and imbuing the field with a fresh vitality.

The aphoristic words of Ludwig Wittgenstein, *The limits of my language mean the limits of my world*, have taken on a new resonance in the era of natural language text-to-image applications powered by AI algorithms. Applications such as Midjourney, Stable Diffusion, and Dall-E 2 have spread like wildfire throughout the architecture community, yielding thousands of stunning images. This explosion of a novel design tool has given rise to two notable outcomes. Firstly, it has produced an abundance of extraordinary images. Secondly, it provokes theoretical inquiries within the architecture discipline, suggesting the dawn of a posthuman design methodology[1]. The confluence of these factors has precipitated a seismic shift within the architectural landscape, fundamentally altering the relationship between the human and the machine in the act of creation. This burgeoning field of natural language text-to-image applications is forging an epic shift in the architectural discourse. By deploying AI-assisted image generators, architects are able to test the waters of the vast ocean of AI without having to bear the burden of coding neural networks from scratch or undergoing the tribulations of creating their own datasets. The advent of ChatGPT has resulted in the obsolescence of promptism and prompt engineering. It has made it feasible for even the most inexperienced user to come up with functional code and complex image prompts. All of this resulted in the proliferation of astonishing images, engendering a new epoch of design that blends human creativity with algorithmic intelligence. In doing so, diffusion models have emerged as a possible new design tool. Enabling architects to mine the multilayered, deep historical repositories of architectural knowledge for chimeras, capriccios, and mutants. Encouraging architects to discover a new voice for the architecture of the 21st century - one that is rooted in bold and visionary experimentation by default. This moment is truly remarkable, not just for the technological innovations revolutionizing the field of architecture, but also for the profound shift in our understanding of design. By working in tandem with AI, architects are not merely creating new images, but rather probing and subverting the very bedrock of traditional design methods. One could even view these image generators as highly advanced accelerators of human ingenuity – expanding the

A House made of Feathers
Midjourney V.3 May 22nd 2022.
It was shown in the exhibition *Strange* in Forth Worth, Texas in June 2022.
SPAN(MdC & MS)

limited possibilities of the mind and allowing humans to peek into the exotic realm of latent space. In doing so, they are opening up new avenues for creativity and experimentation, allowing architecture to transcend the constraints of the past and move into a new era of design. The processing of images through diffusion models breaking them down to mere noise only to rebuild them as something entirely surprising, unexpected and occasionally novel is akin to how architects can deconstruct traditional architectural values and mount them back together into fresh ideas and concepts. Such concepts may better serve the needs of a rapidly evolving 21st century. The proliferation of this new tendency has resulted in a wealth of architectural imagery shared on social media platforms, discussed in the comments sections, and exchanged in informal online meetings. But beyond the excitement and wonder of these images, there lies a deeper discourse that needs to be explored.

Ontology of Image Generators

The ontology of this new architecture, which is based on the use of neural networks trained on pre-existing datasets and pretrained models, has led to an important question about the originality of its outputs. Can a neural network truly create something new when it is built upon existing data? On the other end of the spectrum are, of course, the epistemological questions, which result in a fundamental challenge to the very nature of creativity in architecture. Furthermore, this inquiry goes beyond the question of whether the creations of the neural network are new or not.[2] For all intents and purposes, the question whether the results are new or not might not be relevant at all – as long as it provokes the architect to come up with a novel solution to a design problem. It also probes the extent to which these creations can transform our understanding, methodology, and representation of architecture. Can the transformer's outputs challenge our existing perceptions and practices of architecture, leading us to a new and more profound understanding of the field? This epistemological investigation, therefore, is not limited to the creation of innovative designs but also encompasses the core significance of architecture itself. The very essence of architecture as a discipline and a practice is being questioned, and the answer to this inquiry could have far-reaching implications for the future of architecture. It forces us to think deeply about the meaning of architecture and how we can create meaningful and impactful architectural works in the age of AI. In addition, the dawn of a new era necessitates a shift in the ethical considerations that underpin the discipline of architecture. One of these considerations is the urgent need to adopt more scientific and collective methods of design, as opposed to clinging to the outdated Romantic idea of the solitary genius. Although the concept of the *star architect* has largely faded,[3] it still holds sway over certain territories of the field. It is therefore imperative that we re-examine the meaning of imitation[4] and encourage the ethical sharing of knowledge and ideas. The discipline must also confront the ethical implications of employing a technology that is dependent on the work of countless others. Only by doing so can it be ensured that architecture remains a socially responsible and sustainable practice.

A Linguistic Turn in Architecture

The linguistic turn,[5] represents a shift away from traditional philosophical inquiries and toward an exploration of the role of language in shaping human experience and understanding. This mid-20th century movement was influenced by the works of thinkers such as Wittgenstein,[6] Austin,[7] and Merleau-Ponty.[8] It has fundamentally altered the thinking about language and its impact on everyday live. For the linguistic turn, language is not just a tool for communication, but rather a means by which reality is constructed and how we interpret the world around us. It is through language that we make meaning, and it is through meaning that we understand our place in the world. The impact of the linguistic turn is undeniable, as it continues to shape our understanding of the world and our place in it. The emergence of text-to-image models has given rise to the linguistic turn in architecture, a renewed interest in the power of language and its ability to shape the built environment. At the core of this turn is the understanding that language is not merely a tool for communication but is in fact a fundamental component of architecture itself, one that shapes how we perceive, interact with, and understand the built environment. As such, the linguistic turn in architecture is not simply a technological phenomenon, but a cultural and philosophical one as well. It challenges us to rethink our fundamental assumptions about the nature of creativity, authorship, agency, sensibility, and the relationship between language and the world. By drawing on the work of thinkers like Foucault[9] and Barthe,[10] it invites us to explore the possibilities of a more democratic, collaborative, and open-ended approach to architecture. Together, the linguistic turn and the rise of image generators have created a rich field of inquiry within architecture, one that draws on the insights of thinkers and theorists from a wide range of disciplines. By exploring the complex interplay between language, image, and the built environment, architects and theorists alike are challenging traditional modes of practice and opening up new possibilities for the future of architecture.

The book *Diffusions in Architecture: Artificial Intelligence and Image Generators* presents itself as a collection of images and comments by 25 architects and theorists. It is divided into four large blocks: Suppositions, Commorancies, Vestures, and Estrangements.

Suppositions

Beneath any hypothesis, theory, or model lies a set of silent assumptions known as suppositions. These enigmatic propositions serve as the foundation for our conceptual frameworks, guiding the trajectory of our intellectual pursuits. Like threads woven into a fabric, suppositions shape the texture and structure of our theories, often hidden but undeniably influential. They mark the beginning of our intellectual journey, and their ambiguity and uncertainty reflect the elusive nature of knowledge. While we may draw on evidence and logic, suppositions ultimately rest on our subjective perceptions and intuitions.

Commorancies

Commorancies are a broad collection of architectural entities that can be inhabited, ranging from houses and dwellings to temporary shelters. This archaic term allows for a flexible interpretation of inhabitation, serving as an elastic envelope for the examples presented in this section of the book. The term's morphing abilities also allow it to be translated as *place of residence* or simply *place.* As Edward Casey argued in *The Fate of Place: A Philosophical History*[11], house and home have broader connotations that primarily refer to their spatiality. The strange places depicted in this chapter evoke a sense of spatiality as it would occur in a physical environment, even though they are generated from numerical data. Despite being reduced to images, these places will inevitably spill over into our physical realm. However, a place never becomes merely parasitic in relation to its architectural properties, nor is it merely a by- product of powerful image generators. It retains its own features and fate, its own local being, whether actual or virtual.

Vestures

Within the taxonomy of synthetic imaginations,[12] vestures are unique as they handle the challenge of the exterior, the façade, the frontage, or the wrapping of an object. Vesture[13] is typically associated with clothing or attire, especially as a symbol of rank or status (like coronation robes, regalia, chasubles, dalmatics, kasayas, and so on), as well as with the covering of a specific object or surface, such as a building or piece of furniture. It can also describe the process of putting clothing on someone or something, such as dressing a doll or preparing furniture for a photo shoot.[14] This chapter investigates how diffusion models tend to infuse imagery with moods and atmospheres (analogous to layers of fabric), creating a painterly and cinematic appearance that highlights the vestures of architecture.

Estrangements

The concept of estrangement in architecture exists within the interstices of the recognizable and the defamiliarized, the known and the enigmatic, and the ordinary and the evocative. It is a complex and nuanced phenomenon that eludes easy comprehension, necessitating a deep exploration of the language of architecture its forms, materials, and spaces. The projects in this section live happily in this territory of interrogation. At its core, estrangement serves as a means to challenge assumptions about the world we inhabit. It involves unraveling our perception of architecture (built and unbuilt), pushing beyond the boundaries of our habitual understanding, and allowing us to see the world in a new light.

The explorations presented in this book delve into the intricate interplay between diffusion models, architectural designs, and AI. It is an inquiry that seeks to uncover the nuances of the relationship between human creativity and algorithmic intelligence. By synthesizing these two disparate elements, a novel approach to design is forged, one that is marked by a radical departure from the conventions of the past. This new design paradigm has engendered an entirely new era in which the boundaries between human and synthetic imagination have become increasingly blurred. In this way, the use of diffusion models has ushered in a new level of creativity and experimentation, providing fertile ground for the emergence of hitherto unexplored architectural possibilities.

Suppositions are the silent assumptions that live beneath the surface of any hypothesis, theory, or model. They are the enigmatic propositions that guide the trajectories of our conceptual frameworks and provide a foothold for further interrogation. Suppositions are like threads woven into the fabric of theories, often hidden from plain sight but nonetheless shaping the texture and structure of our intellectual constructs. They are the starting points of our investigations, the launching pads for our journey. In their ambiguity and uncertainty, suppositions are a testament to the elusive nature of knowledge itself.

We may base our suppositions on existing evidence or logic, but ultimately, they are grounded in nothing more than our own subjective perceptions and intuitions. The suppositions in this book are hunches, haruspices, and premonitions of a field in motion. Just as a geologist measures the strength of an earthquake as it unfolds or a volcanologist observes an eruption in action, the texts in this chapter examine the developing impact of diffusion models. Yet, despite their ineffable quality, suppositions remain an essential component of theoretical development. They provide a framework for hypothesis formation, experimentation, and observation. They are the hidden scaffolding upon which our theories are built. Ludwig Wittgenstein argues that suppositions are embedded in every nook and cranny of our language,[1] Michel Foucault posed that suppositions can be challenged and subverted towards critical inquiry[2] while Karl Popper emphasized the crucial role of suppositions in forming a hypothesis.[3] The suppositions presented in the following pages, much like the lines of force and equilibrium in a building, weave a tapestry of thoughts, providing stability and direction in the investigation of diffusion models, guiding the trajectory toward the discovery of new knowledge. Similar to the intricate blueprints of architecture, suppositions furnish a framework for the development of hypotheses and theories, serving as a compass in a journey toward a more profound comprehension of AI in architecture design.

///// SUPPOSITIONS /////

Rice or Pasta? Choose Your AI

Mario Carpo

Many years ago, on a warm, early summer day toward the end of the last millennium, I was visiting my parents in my ancestral hometown at the feet of the Western Alps, in Northern Italy. I was traveling in the company of my then girlfriend, a born and bred New Englander, who was of course much intrigued by the traditions and customs of that old, primitive land. I watched your mother cook risotto last night, she told me, and I am certain she never used any measuring tool. She has no scales in her kitchen; nor measuring cups for flour, grains or liquid. I did not see a thermometer, not even a dial on the oven; nor a timer; and I am pretty certain she never looked at her watch. How can she cook? I have no clue, I replied; evidently, the risotto was edible; why do you not query her, discreetly, tomorrow? And so she did, and it turned out that my mother's reply to every query involving the use of measurable quantities in her cooking (how much of it? to what temperature? for how long...?) was "well, it shows;" or, "you can tell" (apparently, my friend inferred, my mother's telling was based on color, odor, or touch). My friend concluded, while we were on our way back to the airport a few days later: your mother--and, I think she added, this entire country--seem to live in a timeless universe of approximation; the modern world of precision, and the spirit of modern science and technology, based on numbers and quantification, never made it to these shores.

I think that around that time we had just finished our seminar readings on Alexandre Koyré, so we may be forgiven for having momentarily forgotten in that context that the scientific revolution itself was born not far from the small subalpine town where that memorable conversation took place, regardless of Galileo's diet (on which, by the way, much is known). Also in retrospect (but we didn't know it back then), in our modernist critique we were somehow attributing to my mother many of the traits and features that a bit later in time came to define the post-modern, "phenomenological" craftsman, as idealized by Richard Sennet and others ideologues of that ilk, who see the artisan as a man without words and reason, who makes things by feeling and intuition, incapable of explaining to himself--or others--the logic of his doings. All artisan lore is thus seen as "tacit knowledge," an embodied form of technical expertise which can be neither learnt nor taught, if not by some form of mystical empathy (Einfühlung) between sentient souls and more or less animated objects.

In so far as my mother's case is concerned, I suspect we might debunk some of these theories if we could assemble a team of scientists, comprising chemists, physicians, biologists, etc., and ask them to observe her cooking over a consistent duration of time, collecting then averaging and converting their quantitative observations into formal rules that anyone could then follow at will. However, as it happens, we do not need to do that--nor does it appear that anyone ever did--because modern societies have found an easier way to bypass my mother's kind of artisan knowledge. My mother's skills in this anecdote were due to, and inherent in, the food she mostly and primarily processes--rice. Locally grown rice is a natural product, retailed even after milling and polishing pretty much in its natural state--with grains in different sizes and shapes, each with variable, unpredictable contents of starch, to be released during cooking in equally unpredictable ways, which only an experienced cook can detect, interpret, and adjust to as needed.

If you do not have the talent and patience to do that, do as I do--cook dry pasta. Dry pasta is an industrial, factory-made product; its prime component is a blend of different wheats, carefully balanced to obtain steady chemical and physical properties; after mixing with water and other ingredients the dough is steamed, cooked, and dried twice, before and after extrusion and cutting, in controlled conditions of temperature and humidity. In the end, the pasta we get from the grocery is a standardized product--as standard as a bottle of Coca-Cola or a hot-rolled I-beam made of structural steel. This is why its cooking time in a pot of boiling water can be predicted, always the same and the same for all--so long, that is, that the water in which we boil it contains the right amount of salt; in case of decreasing atmospheric pressure one may need to recalculate the cooking time using a formula which I have honed during a life of mountaineering and can forward for free to interested parties on demand. The point is, the entire process can be formalized (converted into rules) and scripted; and when it is scripted, it can be carried out by any person without skill or prior training in the subject--or by any suitable industrial robot, duly programmed.

Artificial intelligence was always meant to do way more than that. In its original formulation (see, for example, Marvin Minsky's seminal Steps Toward an Artificial Intelligence, 1960–1961) artificial intelligence was seen as a "general problem-solving machine" that would find solutions following iterative trial-and-error routines (using strategic shortcuts whenever possible: gradient-based optimization was the smartest and is still largely in use today). In so far as the machine was expected not to repeat the same mistakes, i.e. to learn from its own mistakes, this was not unlike what today we call machine learning--even if the term itself came later. However, when it became clear that the machines of the time were not powerful enough to work that way, this strategy was abandoned and soon replaced by a more practical one, known to this day as "knowledge-based," or "rule-based" artificial intelligence. In this new mode the machine is not supposed to learn anything at all; instead, a selected corpus of human knowledge is instilled and installed in its memory, conveniently translated into a set or arborescence of formalized rules, which the machine will implement step by step when prompted. Architectural historians may remember, as an instance of this logic, Negroponte's 1970 Architecture Machine, which was meant to lead users to design a house via a sequence of if/then queries and multiple-choice options. Human users of the machine were not required to have any knowledge of architecture, because all architectural knowledge deemed necessary to solve that specific set of design problems had been pre-installed in the machine itself. That particular "expert system" famously never worked, but many later and less ambitious ones did and still do. To go back to my anecdote, any such machine, when paired with adequate sensors and mechanical extensions, would be perfectly capable of cooking a perfect serving of pasta. The revival of the original mode of artificial intelligence (now often capitalized and abbreviated as AI) started some 10 or 15 years ago, when it appeared that due to the unprecedented computational power of today's machines (Brute Force computing) and to mere size of today's searchable memories (Big Data), today's computers can indeed solve many unwieldy problems if they simply keep trying. This is how a relatively stupid but massive trial and error strategy came to be seen as key to delivering a functioning artificial intelligence, and indeed the almost miraculous successes of today's AI are due for the most part to machine learning, not to rule-based computation.

The design professions had their first encounters with the surprising image-making potentials of this new breed of born-again AI following the invention of Generative Adversarial Networks in the mid-2010s. As we now know full well, GAN can detect and analyze some commonalities inherent in a consistent visual dataset (a corpus of selected images), extract them, and convert them into a vectorial definition (known as "latent space" in technical parlance). This mathematical abstraction can in turn generate new images similar to those in the original corpus; or can be tasked to combine features taken from that corpus with new images, thus creating hybrids between datasets where, in particular, some formal attributes extracted from one corpus are applied to the content of another. Not surprisingly, this operation was called "style transfer" by the engineers who first described it in 2016;[4] one must also come to the almost inevitable conclusion that GAN can carry out, in this instance, something similar to an intelligent inductive operation which, by dint of comparison, selection, generalization, abstraction, and formalization, automates visual imitation—a cognitive process which has fascinated and eluded philosophers, artists, and scientists since the time of Plato and Aristotle.

The tsunami of Generative AI that has swept the world since the spring of 2022 has not significantly altered this conceptual framework; in fact, for visual artists and designers, the text-to-image stunts of DALL-E etc. are not particularly interesting--at least, not for the time being (even if they are admittedly disturbing, or worse, for linguists and writers). A research paper published in the fall of 2022 by Google's DeepMind[5], which promises to extend the ambit of imitation learning beyond the ambit of styles, pictorial or otherwise, and apply it to knowledge transfer as a whole, may likewise not be of immediate consequence for the visual arts, but it may significantly impact the art of cooking, and--at the opposite end of the hierarchy of intellectual added value--the modern theory of science in its entirety. The DeepMind team, noted for its expertise in imitation learning, which they famously nurtured and honed by teaching computers to learn games, has now trained a robot to learn how to sort a set of colored blocks by observation alone--i.e., by feeding the system a huge dataset showing examples of said sorting, which the computer learned to replicate (i.e., imitate) in the absence of any formal rule telling the system what to do, or how to do it.

This machine, if fed a sufficiently vast dataset--for example, given the possibility to observe my mother's cooking over a lifetime--could learn to replicate my mother's risotto-making skills by imitation alone. It would do so by assimilating, then replicating, the unknown and arguably unknowable rules underpinning my mother's cooking; in true AI style, it would do so without ever spelling out said rules--as, once translated into the system's "latent space" (a multi-dimensional vectorial space), these cooking rules would be perfectly readable, but utterly meaningless to any human intelligence. Since tacit knowledge, as conceptualized by old-school phenomenologists, is mysterious by definition, this mode of AI could then, simply, automate tacit knowledge--sans its magic aura and hocus-pocus.

Even more than the applications, the implications of this mode of machine learning are staggering. Whenever a functional (i.e., big enough) exemplar corpus can be gathered and marshaled for training purposes, we must assume that imitation may replace traditional rule-based procedures. So, for example, an engineer given a structural model of proven functionality but analytically incalculable, may decide to tweak it to adapt it to a different context by imitation alone (so long as the AI system may map each transformation of said model to a "latent space" in turn derived from a sufficiently vast, relevant dataset). This is not unlike what modern science always did--after collecting a corpus of experimental data, sifting, comparing, and generalizing their results; then formalizing rules that can in turn be applied to predict the unfolding of unprecedented but similar events. The only difference is that, this time, we won't do the job, because the machine will do it; and we won't know the rules, because the machine won't tell us. So, in a nutshell, here is the tagline for tomorrow's engineers: when you cannot calculate, imitate.

Is this not what craftsmen always did? This is what computation today does best. But, let's not beat about the bush: this also means the end of modern science--that which came out of the Aristotelian, Scholastic, Galilean tradition; the modern science of which computer science itself is the latest avatar to date. To be precise: this does not mean, literally, the end of that science itself--as we would still need it for many intellectual reasons. It means the end of many of its practical functions. But it's a slippery slope, as some alternative sciences are, as always, lurking in the dark and are ready to jump onto a weakened prey. There is, of course, a viable alternative to all this--toward which I incline. Drop the risotto; stick to pasta.

Taxonomy or the Differences Between Humans and Machines Organizing Data

Matías del Campo

The practice of organizing and categorizing information has been a part of human culture for thousands of years as people have sought to understand and make sense of the world around them. In 1813, Swiss botanist Augustin Pyramus de Candolle coined the term taxonomy by mashing up the Greek words *taxis,* meaning arrangement, and *nomos,* meaning law or science. In a contemporary context, the term *taxonomy* describes the science of classification across various disciplines, including biology, information science, and linguistics. In all these fields, taxonomy refers to the systematic organization of data or information based on rules or principles. Thanks to de Candolle's work, the science of systematization has become a fundamental concept in the study of classification and organization. Taxonomies are hierarchical systems of classification that group similar items together based on shared characteristics, making it easier for humans to find and retrieve information. In other words, taxonomy refers to a hierarchical classification scheme that categorizes things into groups or types. Its applications include organizing and indexing knowledge, such as documents, articles, images, plans, sections, and more, in the form of a library classification system or a search engine taxonomy to enable users to quickly locate desired information. Although many taxonomies adopt a hierarchy, which inherently possesses a tree structure, not all do.

On the following page, you can see the taxonomy of the results from an experiment by Ricardo Guisse, Tam Nguyen, and Spencer Reay. In this experiment, they interrogate how the LAION-5B Dataset is prone to racial and cultural biases. In order to organize their research, they relied on a diagram depicting the taxonomy of their results.

In contrast, the computational method of searching for information involves using algorithms and databases to quickly locate relevant information based on specific search queries. Search algorithms, for example, can be described as computer programs designed to solve a search problem. Search algorithms work by retrieving information stored in a specific data structure or calculated in the search space of a problem domain, where the values can be discrete or continuous. Computational search methods are essential in various applications, such as search engines, machine learning, and robotics. While both approaches have advantages, there are also significant differences between them. They follow the concept of ***don't search, sort.*** This quote is commonly attributed to computer scientist and software engineer Jeff Dean. Dean is a senior fellow at Google and has played a vital role in developing many of the company's most important technologies, including the Google Brain project, which focuses on machine learning and deep neural networks. The idea behind the phrase is that rather than trying to sort and organize vast amounts of data in advance, storing the data in a raw, unorganized form is often more efficient than using search algorithms to locate the information when needed quickly. This approach has been essential for big data applications, where the sheer volume of data makes it impractical to manually sort and categorize everything in advance. This method has also spilled over into the physical world, as demonstrated by the Amazon warehouse approach, where goods are stored randomly, sans the neatly sorted storage commonly used by humans.

Taxonomy of the term *Monument* in Midjourney V4. Thesis Project, Taubman College of Architecture and Urban Planning, University of Michigan. Ricardo Guisse, Tam Nguyen and Spencer Reay 2023

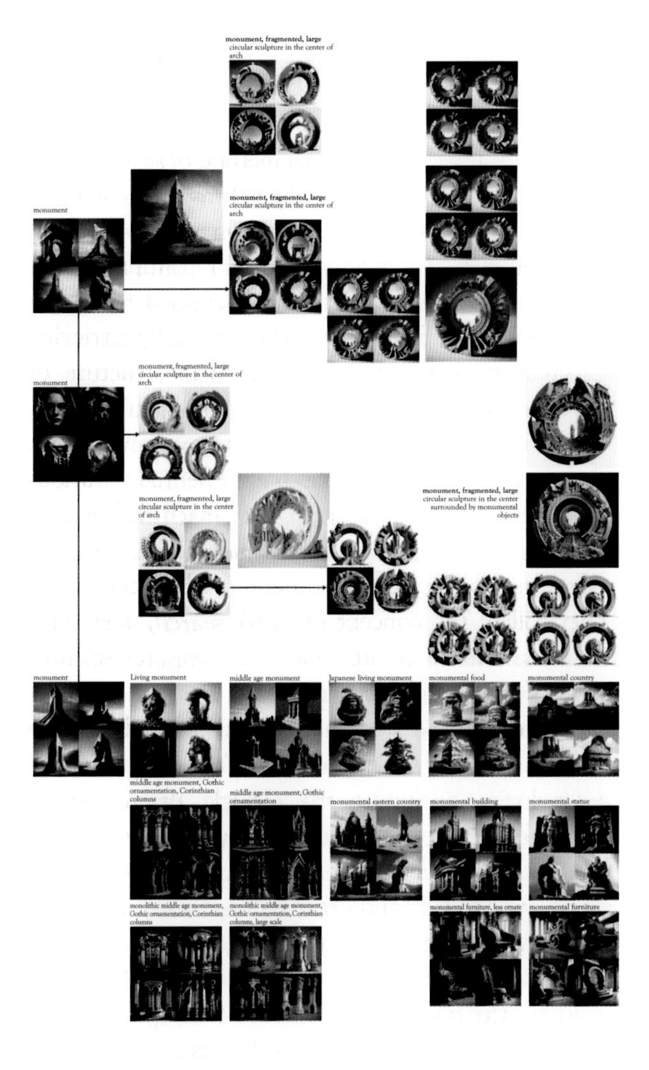

Instead, an algorithm has the overview of what is where and can guide human assistants to retrieve the goods from the shelves in order to bring them to the shipping area. Unfortunately, this method cannot be applied to ink, paper, and glue in the form of the book you are holding in your hand – so, taxonomy it is.

One advantage of the human method of sorting information in the form of taxonomies is that it allows for more nuanced categorization of data. Taxonomies can be tailored to specific fields or domains and incorporate objective and subjective criteria to create more meaningful categories. For example, in the field of biology, taxonomies are used to group organisms based on shared characteristics such as physical features and genetic traits. In the field of library science, taxonomies are used to organize books and other materials based on subject matter and other factors. On the other hand, the computational method of searching for information is often faster and more efficient than the human method of sorting information. Search algorithms can quickly scan vast amounts of data and retrieve relevant information based on specific keywords or phrases. This can be especially useful when dealing with large datasets or searching for information across multiple domains or fields. Another advantage of the computational method of searching for information is that it is often more objective than the human method of sorting information. Personal biases or subjective criteria do not influence search algorithms and can retrieve information based solely on its relevance to the search query. This can be particularly important in fields such as medicine or law, where accuracy and objectivity are paramount. However, the computational method of searching for information also has some limitations. One of the main drawbacks is that it can be challenging to ensure that search algorithms are returning the most relevant results. Because search algorithms are based on specific keywords and phrases, they may miss important information not captured by those search terms. Additionally, the computational method of searching for information may be less effective when dealing with complex or ambiguous data. Of course, a machinic data retrieval process can fail spectacularly when it is supposed to understand the underlying information in the given data, the nuances expressed in a text, or the subtle humorous references, hinted to by a cheeky author.

In this book, I present a taxonomy of synthetic imaginations. Conventionally speaking, this could be a hierarchical system for classifying different types of synthetic or artificially generated images based on their underlying techniques, styles, and applications. However, this book divided the taxonomy into three distinct categories: commorancies, vestures, and estrangements.

Emphasis was given to the notion of the envelope, the wrapper, the vessel, the vesical, and the membrane between interior and exterior. Reaching from a collection of architectural entities that can be inhabited – houses, dwellings, residences, abodes, domiciles, down to temporary shelters – to the cultural and symbolic meaning of vestures. These can be read as frontages, facades, and the dress(ing) of a building. The symbolic meaning of such envelopes is related to the ideas of context and framing. Just as the literary envelope provides context for the main narrative, the architectural envelope offers a frame for the activities within a building. The envelope can shape the user's experience of the building by filtering and framing views of the external environment, controlling the flow of light and air, and contributing to the overall atmosphere and mood of the space – which can be so brilliantly depicted using diffusion models.

In this sense, the envelope can be seen as a metaphorical extension of the building's cultural and social context, reflecting its designers' and users' values and beliefs. The envelope can also serve as a means of communication by conveying meaning and symbolism through its design and materiality. Overall, the envelope is a crucial component of architectural design, playing a vital role in shaping the user's experience of the built environment.

Estrangements, the third big block of this taxonomy, are an animal of their own. They loom large above so many of the results seen emerging out of the primordial soup of synthetic imaginations cooked up by applications such as Midjourney, Dall-E2, and Stable Diffusion. The concept of estrangement in architecture exists within the interstices of the recognizable and the defamiliarized, the known and the enigmatic, and the ordinary and the evocative. It is a complex and nuanced phenomenon that eludes easy comprehension, necessitating a deep exploration of the language of architecture – its forms, materials, and spaces. On a deeper level, the taxonomy of this book reflects the ability of diffusion models to de-structure the data in images into its minute components, reading the RGB values of each pixel and storing its numerical data. Then, using a denoising method restructuring the pixels into novel images full of information that we as humans can read and interpret. So much atmosphere, so many cinematic moments, and so many opportunities for the interpretation of possible meanings. Mind you, machines perform none of these cognitive tasks, but humans are being confronted with results generated by machines – this does not mean that this response was an intention of the machine, but rather that it was inherently present in the data processed by the diffusion model.

The human method of sorting information in the form of taxonomies and the computational process of searching for information have advantages and limitations. Taxonomies provide a more nuanced and flexible way to categorize data, while search algorithms offer a faster and more objective way to retrieve information.

Ultimately, the most effective approach to organizing and accessing information may depend on the specific needs and goals of a particular field or domain.

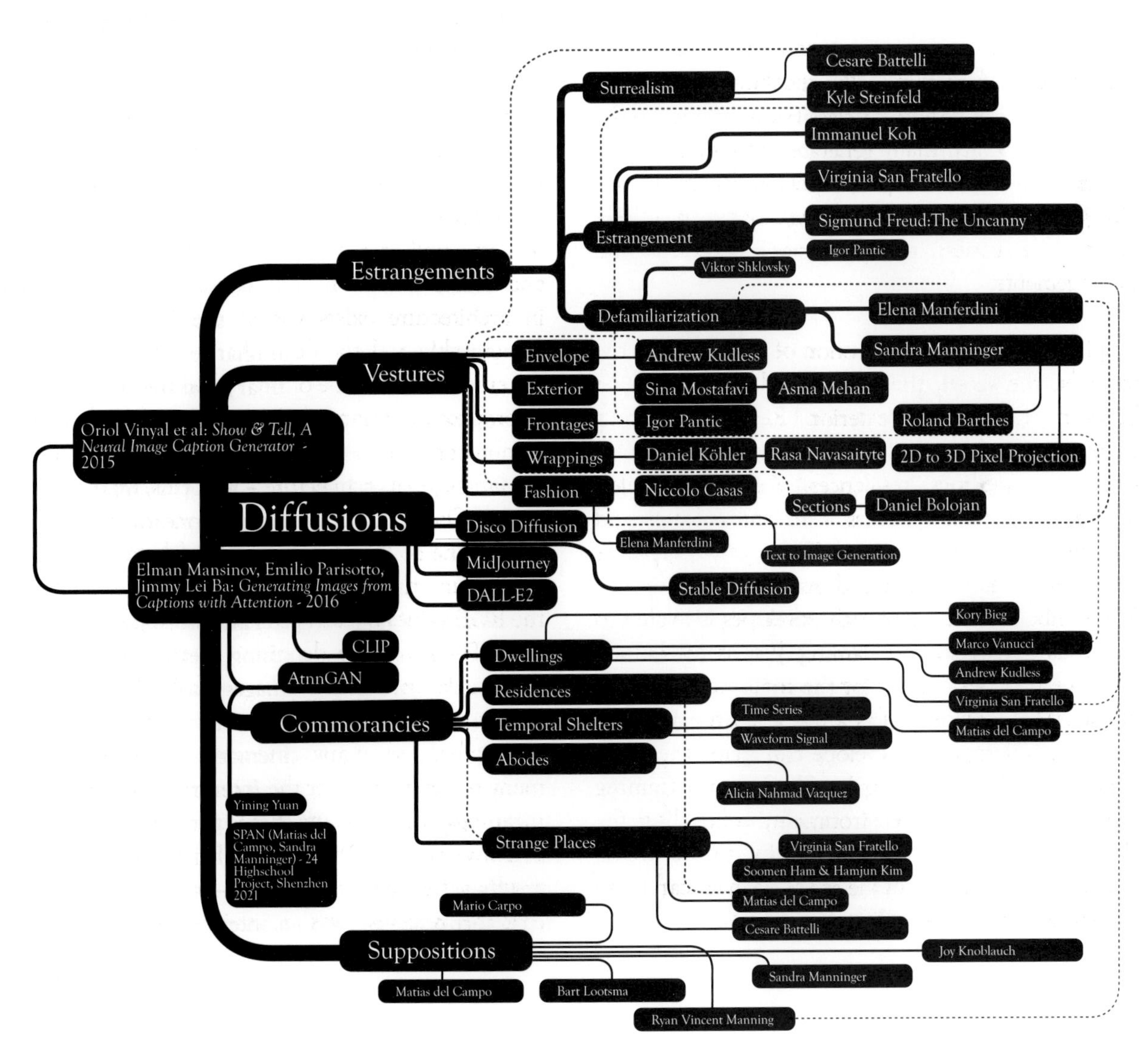

Taxonomy of Diffusions in Architecture Design. Matias del Campo 2023

Diffusions

Bart Lootsma

The real is produced from miniaturized cells, matrices, and memory banks, models of control – and it can be reproduced an indefinite number of times from these. It no longer needs to be rational, because it no longer measures itself against an ideal or negative instance. It is no longer anything but operational. In fact, it is no longer really the real, because no imaginary envelops it anymore. It is a hyperreal, produced from a radiating synthesis of combinatory models in a hyperspace without atmosphere. (Jean Baudrillard, 1981)[6]

The streets no longer lead to fashion's future; today trends break out on the internet. (Dean Kissick, 2013)[7]

Since last year, with the introduction of Stable Diffusion, Midjourney, and Dall.E, several computer programs are available on the internet which, when you enter a text, produce sophisticated images. The images are detailed and show seemingly effortless complex shapes, structures, and textures. Working with these programs is a lot of fun, especially because of the surprising combinations of the known and the unknown the software produces, because you don't know what is really happening, because it is easy delivers results incredibly quickly. The results improve almost every month when measured against their ability to convincingly depict in detail an apparent or possible reality. They will at least produce new aesthetic sensibilities and maybe even more.

Pictorial Turn

Web magazines have surpassed the amounts of subscriptions and page views traditional architectural magazines used to have. ArchDaily boasts 285 million monthly page views, 17,9 million monthly visits, 3.4 million Facebook Fans, and 4.2 million Instagram followers.[8] Not even in their heyday did print magazines reach such numbers. The largest architectural magazines had maybe 30 – 60.000 subscriptions, never more, and today there aren't many left with more than 10.000 subscriptions. The new web magazines have a global reach – both in terms of content and in terms of readers. Whether through photographs or through renderings, images -digital images- are the dominant and most important medium to communicate architecture today already. The sophistication of these images in depicting detail, sharpness, textures, weather, and atmospheres is high and incredibly seductive – whether the project has been realized or not.

In a reaction to the smoothness of most of these images, collage, as a technique to communicate architectural ideas, has celebrated an unexpected comeback. Already in 2013, Pedro Gadanho curated the exhibition Cut 'n' Paste: From Architectural Assemblage to Collage City in the Museum of Modern Art in New York and ever since there has been a steady flow of exhibitions, publications, and symposia around the theme of collage in architecture. Sam Jacob already spoke of a return with a vengeance.[9]

According to Jacob, the comeback of the collage is a return to drawing after rendering software had taken that out of the hands of the architects. It's part of a fight against the alienation new technologies causes.

Growing computational power was harnessed to produce rendered images -glossy visions of soon-to-be-built projects, usually blue-skyed, lush-leafed, and populated by groups of groomed and grinning clip-art figures, where buildings appeared with a polished sheen and lens flares proliferated.[10]

This may very well be true, even if the software Jacob is referring to is also increasingly used to produce staggering dystopian fantasies about, and dark comments on our built-up reality, not least as parts of the special effects industry for moviemaking. This is where this software originally came from before it fell in architect's hands in the famous Paperless Studio in Columbia University in the early 1990s. These fantasies may very well be considered a contemporary equivalent to drawing and painting. The cinematic dystopian architectural and urban worlds of Liam Young, are a kind of homecoming in this respect.

Also, several Italian architects have recently taken up the technique of collage again to communicate their ideas about architecture. The best-known examples are Carmelo Baglivo, Luca Galofaro and Beniamino Servino, but others, like Davide Trabucco, follow in their footsteps.[11] One of the advantages of collages is that one can introduce (parts of) photographic representations of reality, including styles, materials, and textures. But the biggest advantage is that making a collage goes relatively quickly. Even if Baglivo, Galofaro, and Servino also produce books, these collages are posted on social media platforms like Facebook and Instagram in the first place, where they seem to be more at home. They often take the form of memes: combinations of quick remixed graphic ideas with simple texts that work within particular cultural discourses. In the Italian context, one can see different authors communicating with each other through collages. Text plays a minor role here.

Diffusion

It is in this context that artificial Intelligence (AI) programs appear on the scene that can generate equally sophisticated digital imagery as photographs and renderings and produce images even quicker than one could do with a collage. Software like Stable Diffusion, Midjourney, and DALL.E2 can do that for you when you just enter a text. The image appears in less than 60 seconds. The images are detailed and show seemingly effortless complex shapes, structures, styles, and textures. Manifestations of nature – skin, hair, and greenery – are effortlessly depicted. This does not mean diffusion software is very well suited to depict an existing reality. When trying to represent, for example, a portrait of an existing person or city, it only works to a certain extent with examples that are famous in the United States – say, Donald Trump, or Manhattan seen from the Brooklyn Bridge – but they all appear with major faults. Images of less famous people or scenes sometimes only have a faint resemblance to the original.

Even if we speak of AI, we should remember that the current text-to-image models or diffusion models are forms of *machine learning*. They are trained on extremely large datasets of titled images, but these are not simply used as they are. After that, they can remove the noise and *recognize* the prompted image from selected materials.

If we want a realistic depiction of someone or something, we still better go to Google Image Search. We are far removed here from programs like ChatGPT, which – with some reservations – might come close to delivering results that may compete with Google and Wikipedia and in their ability to build up arguments for specific contexts even go beyond their predecessors. At the same time, we should not understand these AI programs as forms of intelligence that can generate something really new and unexpected, but only other orders of existing things. Only every now and then something may go wrong and something unknown may emerge by chance. It is then said that the programs hallucinate.

When working with diffusion models, one tries to understand how one can control the technology after all, but it's not that simple. It's always about the *prompt*: the text that sets everything in motion. But they do more than that: prompts also frame the result and can control the content of the result and its aesthetics to a certain degree. *Prompt engineers* are already very experienced at achieving results that come close to the expected result or even beyond that. Websites like PromptHero show examples and offer courses in formulating prompts to achieve ever more perfect results, whatever perfection may be in this case. If it's about achieving a detailed, seemingly realistic image, that might soon work out. If it's about realizing an image that comes close to an image one has in mind from the beginning, it might remain problematic. But one of the more amusing aspects may very well be that the model produces something unexpected.

Learning and Unlearning

The learning material defines many of the biases of diffusion models. The datasets are provided by firms like the German non-profit organization LAION and the American Common Crawl. The latter collects 3 billion Internet pages per month.

According to The Guardian, *Researchers at LAION took a chunk of the Common Crawl data and pulled out every image with an 'alt' tag, a line or so of text meant to be used to describe images on web pages. After some trimming, links to the original images and the text describing them are released in vast collections: LAION-5B, released in March 2022, contains more than five billion text-image pairs. These images are 'public' images in the broadest sense: any image ever published on the internet may be gathered up into them, with exactly the kind of strange effects one may expect.*[12]

Still, Midjourney's learning material clearly has its focus on American examples. Then comes Europe, and eventually the rest of the world. It affirms the biases of Western society and has clear racial and gender prejudices. If one wants a female professional or a person of color in an image, one must put that into the prompt explicitly.

On top of that, all representations have crucial flaws. It is not without a reason that the first thing you type in the Midjourney bot is /imagine. It produces an imaginary world, a possible world, a proto-surrealist world, the laws of which are the laws of Alfred Jarry's "pataphysics," a physics of the possible beyond metaphysics. It's a world without moral impetus – apart from the biases and censorship introduced by its makers. As such, it's an endless source of creativity.

The resulting images are mainly communicated on the Internet and function similarly as memes in social networks. Intriguingly, Midjourney is even accessed through a social platform, Discord, which was originally developed for online gaming. All images one produces with Midjourney also appear automatically on Discord. They can be downloaded for other purposes from one's own personal homepage at Midjourney.com.

Blind Eyes

The new text-to-image software attracts an immense amount of attention, not just in professional magazines and in universities, but also in the daily press. At the moment of writing this essay, there's hardly a day without an article on AI in the popular press. There are probably countless other ways in which AI is changing and could change our world, some obvious, some more hidden, but it's the strong visual impact of Stable Diffusion, Midjourney, and DALL.E2 and their easy accessibility and use that makes people jump on them. Since Le Corbusier accused architects, they have *eyes that do not see* a hundred years ago, architects do their best to be early adopters of these new technologies. The visual aspect is thereby quintessential. Le Corbusier thought new technologies would mainly change the way buildings would be organized and constructed and, therefore, the way they looked. It was only much later, that new technologies also forced him to change the organization of his office.[13] Today, it seems the other way around. Developments in computerization in architecture since the 1990s have completely changed every architectural practice, even if it's not always obvious if this has changed the way architecture looks – apart from those cases in which architects consciously introduced computation already in the earliest part of the design process. Although there are exceptions that are much celebrated, the inherent conservatism of the building industry still slows down the realization of such projects. Also, the design of such projects is still a decent amount of work for skilled architects. The introduction of software that generates sophisticated images of architectural designs from the start, makes architects willing to speculate about the possible impact of AI on their work to keep up with developments.

There are still some problems to be solved to achieve the results architects are waiting for, notably the current impossibility to relate the imagery to plans and sections. Also, it's not yet possible to insert the AI-generated project in a concrete situation. No doubt there are and will be solutions for that. The fear, that AI will take away work and make many superfluous seems to have vanished.

Text Prompts

One of the most intriguing aspects of text-to-image software is the new relation between the two. The prompt is a shorter or longer text. It's a command that generates the image, no longer a description of an image that is already there. Similar phenomena play a role in illustrating and in conceptual art. Of course, illustrations to a scientific text or to a manual are supposed to be as precise as possible, but those made for a newspaper article, children's books, or comics are much more open to the personal interpretation of the artist. This seems one of the most promising fields in which text-to-image software can find a use. Cartoons and caricatures, which exaggerate a certain situation, are other options.

In art, the title or description is usually added after the visual work has been realized. The idea is, that the visual work speaks for itself, even if that's not necessarily the case. From the end of the 19th century and particularly in the 20th century, the title and even longer texts relating to the visual work became more important. In conceptual art, the complex relationship between image and text has became a recurring issue. This is already evident in the work of Marcel Duchamp, who changed the meaning of everyday objects by putting them in an art context and adding a title, often in the form of a pun. Duchamp's Green Box from 1934 is already more ambivalent, as it contains notes and sketches related to his magnum opus The Bride Stripped Bare by Her Bachelors, Even, or Large Glass, on which he worked between 1915 and 1923. Some of the notes and sketches in the Green Box anticipate parts of the Large Glass; some are facsimiles; some describe or depict parts of the Large Glass that were never realized; and some relate to other works and thus embed the Large Glass in an even larger universe. The combination of the Large Glass and the Green Box produces a complex world of ideas, open to different interpretations. But Duchamp also collected his puns as works in themselves, published them, and recorded a spoken version of them, triggering the imagination of the audience in another way. In the 1960s and 1970s, artists as different as Joseph Kosuth, Robert Barry, Lawrence Weiner, Marcel Broothaers, Sol Lewitt, Joseph Beuys, and many others would produce works that would either just consist of texts or texts by means of which someone else could realize a work – maybe even in different contexts. In his book The Second Digital Turn, Mario Carpo reminds us that before the renaissance, *the main vehicle for recording and transmission of visual data was verbal, not visual: images were described using words; written words were forwarded in space and time, images were not.* And he refers to Isidore of Seville, who epitomized the ancient mistrust of all forms of visual communication, and stated that *images are always deceitful, never reliable, and never true to reality.*[14] If it is true that, as Carpo writes, *the rapid progress of contemporary digital technologies from verbal to visual to spatial media in the course of the last thirty years curiously reenacts, in a telescoped timeline, the entire development of Western cultural technologies* many of these issues will probably be solved.[15]

I notice in my own experiments that some people can now be made to believe that the results are photos when I post them on Facebook or Instagram. For example, there's a series in which I prompted Midjourney to generate young versions of famous architects, with attributes that are semi-related to certain familiar narratives associated with them. Most people, of course, do not know what these people looked like when they were young. Nevertheless, many accept the suggestion. Mostly, one finds only vague hints of the real persons in them, as many or as few as if one had a current or timeless portrait made. The only difference is that people accept it more when they are portrayed *young* in a photorealistic way, because most people looked different when they were young. The other way around, I notice that people start doubting real photos when I post them after Midjourney images. This is understandable, as many of these images have been photoshopped before they were posted or printed. This anticipates some of Midjourney's aesthetic biases and has prepared us to accept them.

Acceptance may have a lot to do with the speed and superficiality of these media, with the textual descriptions, and not least with what people want to see or accept as true. The role of the descriptions is central here: they are not added later, but they are the origins of the images. In this way, they also challenge us to see the images as realizations of these commands. At the same time, Midjourney makes it clear that not everything is to be understood as text and that a linguistic summary of a reality or idea is always a simplification. The images are much richer in information than the prompts.

Guilty Pleasures

The deceitfulness and unreliability of Midjourney images are also inherent to the very essence of diffusion models. They feed on the Internet and feed the Internet themselves in an incestuous process. It's all Simulacra and Simulations, as Jean Baudrillard would say. He wrote already in 1994 that *today abstraction is no longer that of the map, the double, the mirror, or the concept. Simulation is no longer that of a territory, a referential being, or a substance. It is the generation by models of a real without origin or reality: a hyperreal.*[16] In the case of text-to-image software there may be millions or even billions of origins to the produced image, but these are all blurred and deconstructed. Baudrillard defines the successive phases of the image as: first, the reflection of a profound reality; second, the masking and denaturization of the image; third, the masking and denaturization of a profound reality; fourth, the masking of the absence of a profound reality; and finally, the phase in which the image has no relation to any reality whatsoever and becomes its own pure simulacrum. This is obviously the phase we have reached now.[17]

By far the majority of images generated by the new AI programs belong unmistakably to the categories of fantasy, science fiction, and horror, including the scary psychedelic colors that go with them. Areas, in other words, that traditionally already consist of a mixture of exaggerated realism, historical references, and complete nonsense. As Roland Barthes already wrote about Martians, the whole of psychosis is based on the myth of the identical, the double.[18] This is more than satisfied by Midjourney. Midjourney's strength, its incredible detail and richness of texture, also becomes a weakness here. The images generated, precisely because of the abundance of clichés, details, textures, and moods, inevitably become kitsch. And according to Umberto Eco, kitsch is *the ideal food for an indolent public that wants to access and enjoy beauty without having to try too hard.*[19]

Does this mean that Midjourney is fundamentally useless? On the contrary. We are only at the beginning, even if, as the name suggests, we are in the middle of the journey. And this journey is as fascinating as it is dangerous. The best thing to do, instead of having Martians designed, is to enter this world as Martians ourselves, like a foreign planet on which we try to get along in all innocence. I suppose many people who enjoy working with Midjourney know it is a hyperreal world of simulacra, and they know that a large part of the production is kitsch. They consider this tongue-in-cheek as a guilty pleasure, in other words: as a form of Camp.

According to Susan Sontag, Camp is a style that is ironic, theatrical, and exaggerated, characterized by a love of the unnatural, artifice, and the artificial. She argues that Camp is a way of seeing things that goes beyond mere style or taste and that it involves a certain degree of aestheticism and frivolity. She also notes that Camp is closely related to the concept of *bad taste* and that it often involves an appreciation for things that are traditionally considered

Susan Sontag in a cave-like Art Nouvea

low or vulgar. In fact, many of the examples of Camp Sontag gives in her famous essay could be Midjourney favorites. Under version 3, results often combined a kind of impressionist painting style from around 1900 with a preference for Art Nouveau-like forms. Sontag calls Art Nouveau as the most typical and fully developed Camp style. *Art Nouveau objects, typically, convert one thing into something else: the lighting fixtures in the form of flowering plants, the living room which is really a grotto. A remarkable example: the Paris Metro entrances designed by Hector Guimard in the late 1890s in the shape of cast-iron orchid stalks.*[20] Sontag argues that camp is often most effective when it appropriates elements of low culture, transforming them into something that is both ridiculous and sublime. In most cases, this is exactly what the diffusion models do. Sontag sees camp also as a mode of cultural production that is both celebratory and critical, a way of embracing and reveling in the absurdity and excess of modern life while simultaneously exposing the artifice and artificiality that underlie it.

The enormous impact of text-to-image models will probably change aesthetic sensibilities in architecture and design. And maybe someday we can design with this confusing new AI-infused software and thus project it back into reality. After all, in the early 1990s, when special effects software like Maya only ran on extremely expensive Silicon Graphics machines, what is now normal in everyday use could not be expected immediately either. And this development is accelerating. *The streets no longer lead to fashion's future; today trends break out on the internet*, wrote Dean Kissick of the fashion magazine i-D. The same will be true for architecture, design, and probably the whole of visual culture.[21]

The Labors of AI

Sandra Manninger

An in-Depth interrogation of Labor in the context of dataset creation for Diffusion Models

Have you ever considered how data comes to be? Or, have you ever thought how architecture can avoid the pitfalls that happened historically in the creation of datasets? A crucial component so urgently needed to contribute qualitatively, inclusively, and ethically to current developments in architecture design. In this text, I try to unfold the ontology of data (how does it come to be?), and the epistemology of datasets in this world, where datasets and machine learning (ML) applications increasingly define how traffic flows, goods are delivered, bank loans are decided, stock is traded, and parole is given. Does data appear out of thin air? Does data emerge in a spontaneous generation, perhaps? Aristotle was wrong in his belief that life can spontaneously emerge from inanimate matter[22] – the same as data does not emerge out of nowhere. The creation of these datasets involves labor. Human effort. Work. Like every human work, it involves flows of capital, cultural biases, and questions of ethics. For the architecture discipline, current tech companies can be described as anti-role models. Why? Current artificial intelligence (AI) tech companies rely heavily on surveilled workers such as content moderators, data labelers, shared ride drivers, warehouse stowers, pickers, and packers. Some startups are even hiring people to pretend to be an AI systems (!) like chatbots, as venture capitalists put pressure on software developers to include AI in their products.[23]

Labeling Farm, according to Midjourney. Prompt: Photography of an MTurk Digital Sweatshop, people of diverse races Labeling Images -v 4

MMC Ventures, a London-based venture capital firm, conducted a survey of 2,830 AI startups in the European Union (EU).[24] The result was that 40% of these startups did not use AI in a meaningful way. Despite representations in popular culture, AI is far from achieving sentience. The underbelly of current ML and machine vision applications consists of millions of underpaid workers around the world. Their work consists of mind-numbing repetitive acts, executed under questionable labor conditions. Literally piecework labor. In contrast to the white-collar workers of Silicon Valley, who receive six-figure salaries and more, the low end of the work population in the labeling industry is habitually recruited from impoverished segments of the population and is paid as little as $1.46/hour after tax in digital sweatshops.[25] Often, they get paid per labeled image – a couple of cents per image.[26] The more surprising is that the discussion of labor conditions regarding the creation of Datasets – the basis of any AI application– is not in the center of current conversations about the ethical development and proliferation of AI systems. Even less so in architecture circles, where debates more often than not gravitate around algorithms and models. As architecture is a relatively new field when it comes to the integration of AI applications into their workflows, now is the right time to discuss the exploitation of workers that makes the application of AI in architecture possible. As the discipline moves to create its own datasets that will inform architecture possibly for decades, it is the right time to plan how to avoid the exploitative techniques applied by large social media conglomerates and small start-ups alike in recent times. In this essay I attempt to dive deeper into the multiple layers of labor conditions related to tasks related to ML and other AI applications to demonstrate current working conditions. Including considerations pertaining to worker organizations within the architecture discipline regarding labor like labeling and annotation should be central in debates about AI, ethics, and architecture.

The Latent Space of Labor: Hidden but Visible

As the architecture discipline is on the verge of creating large-scale datasets in order to train ML processes with a high degree of accuracy, I will rely on the creation of another benchmark-setting dataset to explain, with the aid of historical analysis, the genesis and the involved labor in creating datasets: ImageNet. There are larger, and newer examples, such as the LAION 5-B dataset, but the goal is to rely on the well-documented historical account of the creation of a dataset in the context

The vast majority of crowdworkers are situated in three regions of the world: India, Europe, and North America according to Midjourney. Prompt: photograph with flash of protesting digital workers, members of the diverse cultural background, Computational Proletariat, crowdworkers –v 4

of labor. This dataset is one of the standards today when it comes to machine vision tasks such as the ones used in object recognition,[27] image classification,[28] and object localization.[29] (Sidenote[1]: It might not be possible to achieve the scale of datasets like ImageNet with all the architecture images in the world, so it might be necessary to synthetically augment the dataset, but more about this topic will be discussed later. Sidenote 2: Even modern labeling methods such as zero-shot learning rely on predefined, structured descriptions that include labeled meaning and auxiliary information, or as Parmenides put it so eloquently, nothing comes from nothing.)

Fei-Fei Li started the work on ImageNet in 2006, when most AI researchers were actually focusing on the construction of algorithms and models. The similarities with current debates in architecture are uncanny! Li, however, recognized early on that the quality and especially the amount of data points are crucial for the robustness of the results of any AI process. (provided it uses images, of course). The origin point of ImageNet can be found in a meeting between Fei-Fei Li and Princeton Professor Christiane Fellbaum, who was working on WordNet[30] at the time. A popular dataset for applications such as information retrieval,[31] automatic text classification,[32] automatic text summarization,[33] machine translation,[34] and automatic crossword puzzle generation.[35] Li used the basic frame and many of the features present in WordNet as a template for ImageNet. Instead of focusing on the refinement of an algorithm that at some point could create robust results regardless of the data, Fei focused on mapping the world in order to provide more examples to learn from for ML processes. However, there was a huge obstacle. ImageNet demanded an enormous amount of annotated data that went far beyond the scope of the work and the financial possibilities of her laboratory. According to Fei-Fei Li, one of the challenges of constructing ImageNet was the verification of the vast amounts of data gathered from Internet search engines. At first, Li considered hiring undergraduate students to do the labeling work. $10 an hour in order to manually retrieve images, add them to the dataset, and annotate the content. Simple math revealed that using this approach, it would take 90 years to come close to any essential number of annotated images.[36] Furthermore, undergraduate work is subject to interruptions as it depends on external factors, such as the time of the school

Allegory of Fei-Fei Li at the moment she conceived the idea to ImageNet. According to Midjourney. Prompt: photography of Fei Fei Li inventing ImageNet --v 4

year, funding, and training. Furthermore, such a contained population of labelers would without question introduce biases into the dataset. For example, just to mention one, in terms of racial bias. The current statistics[37] show that the student population at Princeton consists of 39% White; 28% Asian; 11% Hispanic, Latinx, Mexican American or Puerto Rican; 10% Black/African American; 6% accounts for two or more races, Native American, Native Alaskan, Native Hawaiian, and other Pacific Islander; and 4% is unknown.

The idea of labeling with the help of Princeton students was abandoned quickly, and alternative routes were explored, such as using automated systems for image retrieval and annotation and applying for federal grants to finance the dataset building. After abandoning the idea of undergraduate-enabled labeling due to the enormous costs that would have been involved, the ImageNet group attempted a computational method of labeling, in which an algorithm would sort through the mountains of data and reduce the overall human labeling effort. This approach proved to be erroneous, as the quality of the labels would be limited by the capabilities of the machines at the time of construction. This, as Li describes, would contradict the stated goal of constructing a *gold standard* dataset,[38] which that should ultimately be set by humans. All of these attempts failed: *algorithms were flawed, and the team didn't have money.... The project failed to win any of the federal grants she applied for, receiving comments on proposals that it was shameful that Princeton would research this topic and that the only strength of the proposal was that Li was a woman.*[39]

The *deus ex machina* in this story was a student. In a chance conversation in the corridors of Princeton. A student asked Fei-Fei Li whether she had heard about a new service provided by Amazon called *Mechanical Turk*. The basic idea behind it is the crowd-sourcing of labor, doing simple tasks in a distributed fashion. For Pennies.

He showed me the website and I can tell you literally that day I knew the ImageNet project was going to happen.[40]

The name of the service "Mechanical Turk" has its origin in an 18th-century automaton that was built in Vienna, Austria. In 1771, Wolfgang Ritter von Kempelen presented the chess playing automaton *The Turk* to empress Maria Theresia at Schönbrunn Palace in Vienna, Austria.[41] The Automaton astonished the audience by expertly playing chess and beating luminaries such as Benjamin Franklin and Catherine the Great. Napoleon Bonaparte cheated in his game with the Turk, but the machine just wiped all figures off the board, after Napoleon's cheeky attempt to cheat the Turk. The automaton consisted of a large plinth in the shape of a trunk with the chessboard on top and the Turk sitting behind it. Kempelen demonstrated the machine before every game, opening the trunk to reveal a complex set of struts and wheels. However, this automaton was fake. The cogs and wheels had a mirror as background, creating the illusion of depth; however, there was a person hidden in the trunk that moved the chess pieces. This story serves as a metaphor for the current methodology for creating datasets. Ironically, the Mechanical Turk, a service

Ritter von Kempelen's Mechanical Turk, 1771, image: Joseph Racknitz, 1789

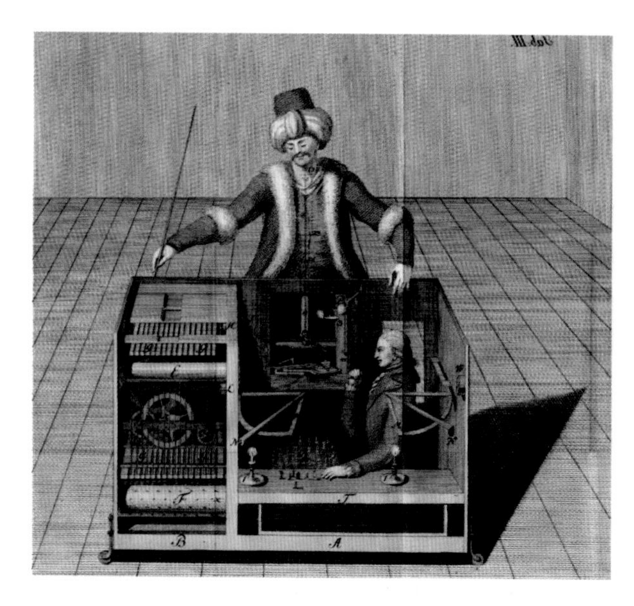

by Amazon Web Services with the same name, seems to pretend to be an automaton but, in fact, it is people annotating images manually.

Circling back to the creation of ImageNet, what is of note is how these two failed labeling attempts reveal how the ImageNet creators conceptualized the labeling task. It seems that in their thinking, humans, rather than machines, are considered critical in the process of obtaining high-quality labels, the constraints on which humans should define the *gold standard* are posed largely in terms of cost and time efficiency. Put differently, the creators of ImageNet set out to install a techno-social configuration which would place humans in the position to speedily perform basic tasks of image recognition without interruption and at a low cost. Their choice to use the Amazon Mechanical Turk (AMT) crowd-working platform was, according to Li, a "tool that could scale, that we could not possibly dream of hiring Princeton undergrads."[42] On this new platform, anyone could construct a "human intelligence task" to be completed by the platform workers, who would be paid for each item they completed. This solution quickly solved their image annotation problem by allowing the problem to be broken down and distributed to 49,000 workers from 167 countries.

Amazon's Mechanical Turk was actually launched the year before Li came up with the idea to create ImageNet: November 2, 2005.[43] In the wake of its launch, the Mechanical Turk user base grew rapidly. Tens of thousands of jobs were posted to MTurk already from early to mid-November 2005, all of which related to Amazon itself, primarily for internal tasks that needed a certain amount of human cognitive abilities. Those included tasks such as image labeling, surveys, rating, transcribing, and writing. By March 2007, the worker base at MTurk had reached more than 100,000 individuals from more than 100 countries.[44] By January 2011, this number had inflated to more than 500,000 registered MTurk workers from more than 190 countries.[45] Also in 2011, Techlist put together a map that allowed one to pinpoint the locations of 50,000 of their own MTurk workers around the world.[46] Despite the high number, research conducted in 2018 disclosed that of the over 100,000 workers subscribing to work on Mechanical Turk, only about 2000 can be considered active accounts working on tasks such as labeling and annotations. In the same year, Techlist published an interactive map that pinpointed the locations of

50,000 of their MTurk workers around the world.[47]

Giving Credit in the Wild West of Data Acquisition

The creation of ImageNet revealed another flaw in the system. In some descriptions, data is treated like a natural resource; its existence only acknowledged as something that can be mined for the benefit of the person extracting the value out of the raw material. Data is the new Oil,[48] is a well-known axiom that poignantly describes the notion to think of data as raw material. Raw oil needs to go through a refinement process to be converted into gas, plastics, medicine, etc. This process converts raw oil into the commodities we desire. Data behaves in a similar fashion, without processing, data is inert – a pile of numbers with no value.[49] The same method of thinking applies to many other datasets such as LAION 5B, COCO, Kinetics-700, IMDB-wiki, and many others. However, data is not a natural resource, it is a product of human activity, consciously created or deliberately harvested by individuals who deposit the data in large repositories designed to aid in ML processes. The creators of ImageNet created a precedent, with after effects to this very day, by ignoring the fact that MTurk is a crowd-working platform. The consequence is that the people collecting the data remain anonymous. This is not something built into the MTurk system, but rather a deliberate decision of the team involved in the conceptualization of the ImageNet dataset.[50] The consequence is that the workers collecting the data get no credit whatsoever for contributing to the dataset. Ethically speaking, a more than dubious methodology. The creators of ImageNet never disclosed how much people were paid to label images. Neither do they discuss which countries had the largest number of annotators nor disclose any demographic characteristics of their annotators. This lack of information is structural, designed to suppress the idea that there is an individual contribution in the dataset – the anonymization of labor.

Designed to be perceived as a generic human intelligence resource with the ability to execute the required image-labeling tasks on the AMT platform. The common ground of this idea is that humans generally have the innate ability to recognize images in the same way–an approach to vision that erases lived experience from the formation of meaning.[51] If humans are used to include semantic information in datasets, should they not be credited for this work?

Lithium mining in the Atacama Desert, Chile (according to Midjourney) Prompt: photography of the Lithium Mine in the Atacama Desert in Chile, cinematic illumination, National Geographic magazine photography--v 4

Common House Dataset. Web interface for crowd sourcing of plans

Common House
JULY 19, 2021 / PROJECTS
Common House project received Arts Integrative Interdisciplinary (AIIR) Faculty Grant from Arts Engine.
The project focuses at the problem of floor plan analysis and generation using neural networks.
An example from the annotated data created.
The main obstacle that has emerged in creating more real-life plans is the lack of databases that are tailored for these architecture applications. The Common House project aims at creating a large-scale dataset for plans with semantic information. Precisely, our data creation pipeline consists of annotating different components of a floor plan, for e.g. Dining Room, Kitchen, Bed Room, etc.

The inclusion of at least a small set of information points such as the location and ethnicity of the MTurk workers would open up a vast amount of additional information that can be the basis for several data analysis tools that allow for improved the performance of the intended ML process. Of course, this would need to be on a voluntary basis, so as not to extract more information for free from workers. A difficult line to walk with respect to ethical criteria for data collection.[52]

A counter example could be how data are collected in dataset projects such as the Common House and Model Mine.[53] Both of these datasets do indeed give credit to every single annotator, making sure that their voices are heard in the data collection effort, helping to understand the architectural problem not only from a global perspective, but also on a regional level, as it is possible to pinpoint the locations of the annotated plans or 3D models. An important aspect in setting data collection efforts such as Common House and Model Mine, is the focus on a diverse population labeling the data. A combination of a diverse student body, with a crowd-sourcing effort from around the world, allowed us to create a diverse dataset of apartment plans from all over the world. In a second stage, annotators were also allowed to upload plans that they considered interesting for the dataset. As mentioned earlier, there is also the problem of the dataset scale, in particular, in architecture. Despite the fact that the discipline of architecture has more than 3000 years of examples under its belt, it is indeed a data-poor discipline. The reason for that is the lack of annotated data, but more importantly, the lack of a tradition of sharing information. Architecture might be the only discipline on the planet that still believes in the sole genius – the result of this disciplinary position results in a series of secretive methods. This notion is a huge obstacle in achieving architecture that is truly responsive, and allows for elevating the living conditions of millions, improving the environmental performance, low material consumption, structural optimization, mindful social integration, cultural diversity, and so on and so forth.

Foucault, Archaeology, and Dataset

The project excavating.ai that was launched by Kate Crawford and Trevor Paglen[54] provides the opportunity to ponder on questions regarding dataset construction, in particular in the light of morally dubious labels. The focus of their research is on the person subcategory of the ImageNet hierarchy that contains labels such as Call Girl,

Drug Addict, Closet Queen, Convict, and more. Crawford and Paglen describe this project as an archaeology of datasets. In the work of SPAN concepts of archeology as defined by Michel Foucault were interrogated for their ability to provide a frame of discussion to explore the modus operandi of Neural Networks (NN's). In particular, with regard to their ability to process historical information to perform salient feature recognition and interpolated image generation. Archeology, according to Michel Foucault, is generally defined by the positionality of statements, images, and discourses in order to specify the boundaries or limits between what can be done, said or considered within a given epistemological and historical context. As exemplified in this essay with the historic example of the creation of ImageNet. In the case of Crawfords's and Paglen's concept, it is framed by a political archeology that reveals the layers of values, motivations, and assumptions that represent patterns of meaning which inhabit datasets and can lead to the distortion and subversion while using the dataset. The goal of this archaeology maintains that by exposing the inherent biases in the dataset, it is possible to directly influence the dataset creation process and exorcise, at least to a certain extent, the inherent labeling and annotation processes.[55] In the bigger picture regarding dataset creation, and in light of the emerging field of architectural datasets, it might be relevant to discuss alternative interpretive methodologies.

Aaron Plasek, for example, maintains that "we need to write histories of the datasets themselves'.[56] This would allow one to critically interrogate the methods applied in the formation, alteration, maintenance, and application of the dataset. This is, however, not reduced to the data alone; this elusive entity sitting somewhere in the cloud, including subjective elements such as assumptions, motivations, and values, as well as the objective and material aspects of dataset infrastructure creation and maintenance. This, of course, is directly connected to the actual extraction of material from the Earth. The lithium, tantalum, palladium, gold, and silicate are necessary to build the hardware that houses the datasets. Similarly to MTurkers, literal miners involved in the extraction of rare earths for the production of the GPU's necessary to apply data to any AI application form a part of the working population that allows the application of ML solutions on a planetary scale.

Foucault's[57] idea of genealogy relies on an interpretive method, identifying and tracing aspects of emergence, formation, and transformation of discourses, concepts, and methods inherent in particular historical contexts. The goal is to identify material conditions and variations in modes of subjection. To do so, he uses a set of resistance strategies that are occasionally divorced from the explicit intentions of the actor. For example, by introducing irreducible forms of resistance, where dominant practices collide with minor discourses. Genealogies, in this frame of thinking, suggest that the modes of power of specific actors allow only actions following a particular conditioning.[58] As Emily Denton describes it, the genealogy of data is not reducible to excavating the hermeneutic ground, that is, the hidden meanings related to the values,

Annotators protesting for their rights, according to Midjourney. Prompt: photograph with flash of Digital Computer white collar Union Workers protesting on the street, holding their laptops in the air, members of the Filipino Computational Proletariat, men and women, --v 4

assumptions, and motivations of particular actors, nor is it reducible to analyzing discourses and practices from a relation between signified and signifier.[59] Instead, the focus of genealogical analysis is on the strategic emergence of discursive events operative in a given dataset infrastructure. Archaeology and genealogy are each in their own way, an attempt to reconstitute historical events. They seek to uncover events verified from their own respective spheres of jurisdiction: archaeology as defined by the historical regularity of the sayable and its limits revealed by epistemological modes of exclusion; genealogy constitutes knowledge within a historical field of power relations, examining the various modes at play in data subjective formation.[60]

To that end, the formation of subject positions and data roles and their deployment in a network of power relations have resulted in the formation of the conditions of possibility, such as the technological affordances of novel crowd-working platforms. These power relations will define the future of work more than we currently can imagine, or even perceive.

Digital Unions

The conclusion of all of this is that Digital Workers, MTurkers, Labelers, Annotators, and Flesh Algorithms need to organize themselves, demand their share in this new economical system that is spreading around the planet like wildfire, and make their voices heard. How does architecture fit into this new ecology of work? As the discipline embarks on the creation of their own datasets, which will define the way we improve architecture in the upcoming decades, the discipline needs this information as it is more than likely that image annotation and labeling will become part of the portfolio of near future architecture offices. Thus, the discipline needs to be equipped with the respective knowledge base of data extraction, labeling, and the ways how today's methods try to be exploitative. The proliferation of the information in this essay might help to raise awareness within the discipline of architecture that there are two major aspects that need to be considered when entering the annotation game: bias and labor justice. These problems need to be tackled by the discipline today, before either it is too late, or someone else tries to solve them for us architects – without our voice being heard.

An Artificial Tale

Ryan Vincent Manning

The first artificial intelligence (AIs), the artificial precursors, appeared here and there via small exclusive groups, where they attracted academics, game enthusiasts, and artists. The pixel-to-pixel and image transfers, witnessed only by a handful of outsiders, were dismissed as computational follies. Even early news reports did not quite know what tone to take, shifting reluctantly from technical descriptions to ironic jests with armored skepticism. Processing power was restricted due to a lack of universal blockchain and optical schematics. Databases requiring large amounts of data only producing moderate results, were limited to research grants and aerospace institutes. The *do-it-yourself* databases were hampered by their size, as generating sufficient data was a laborious task. Web scraping, a method of data collection, whirled around misuse of data, such as polarized politics contorting Pokémon morphologies. One could only collect so many planes, trains, and automobiles before the pot would run dry. Practitioners could gather data only in the thousands, while millions were needed.

At this time, according to Sir Rehms, AIs were more widespread than initially believed. The relentless tides of capitalism, neo-liberal-accelerated science, gradually smacked us with their proverbial hand. In the first wave, lockdowns, social stops occurred. WP I or world pandemic one, as we call it today. The draining social aptitude propelled attentions to early Ais. Stay at home! – old crows screamed – the rest lost in confusion and dismay. Social enclosure. A zero degree of entertainment. Brought AIs to the forefront. Serious flaws, and the time needed to build large databases, became less important as people had all the time in the world. Writers, artists, and architects embraced AIs and allowed them to feed their cave-person solitude. DALL-E, Midjourney, and Disco became in-house names, and image production became the neo-social pastime. Disco Diffusion was my entry drug. Liquid Hejdukian landscapes inhabited my screen, my Instagram, and my tweets. Directions and designs that once took weeks, months, or even years to develop are now at our fingertips. Whales, bears, and bacon-smeared water colored utopias with every click. Conversations about the new technology infested Zoom meetings, conferences, and daily workouts. Our shiny new toy seemed to brighten our lingering depraved loneliness, repeated walks around the neighborhood, and streaming television series. *There is nothing like the first time.*

Google animal dreams and Disco's ink blots were only the beginning. Shortly thereafter, DALL-E 2.0 birthed new utopias and landscapes. Back then, everyone needed an invite. Do you remember those exclusive parties where you had to call to find out the secret location? The private talks, the lunchroom for the lucky few? The Instagram, the Facebook, and the Twitter became the neo-battleground for *Who's got it? Did you see my whale series? It was done in Adam 3.7. No, no you must try Shasta 1.5!* The directional masquerade melted away via prompts and text. This is how we did it: on the computer, mostly with game-oriented chatbots. At first, we were inclined to write Gonzo Keroacian prose; this would never work. One would reduce one's interest to four or five words: thing, where, style. A neo-modernism literature via text message aesthetics or James Bond drink

C

orders in short explanations. Art began to become democratized, or at least that is how we figured it. A big fan of Van Gogh? Just prompt it! A myriad progressed, displaying sunflowers, crabs, and blue, starry nights. Abstraction could now have tea with figurative shapes in a Japanese garden. Michelangelo found Prufrock and they drank coffee, explaining their own exploits sexy and hellish outcomes while watching the passersby. One friend proclaimed, in a grandiose manner, Dyson should hire him for a series of Geiger home appliances, the alien vacuum and the grotesque curling iron. The excitement was short lived.

Bye bye American Pie! We gradually saw the music die, and become a little tarnished, a little less. Midjourney, the new cat on the scene, went public on TMZ. The once exclusive lunchroom was now open to all, as a swarm of designers, artists, and architects saturated the electro web. Stories and blogs generated posts after posts of this neo-art. Everyone became new; everyone became contemporary. *We have now gone Hollywood. We are all influencers now!* The news reports came out and we knew, what used to be so unique was now everywhere. There were, moreover, serious flaws in these early AIs, which became crystal clear. The concept artists, shaking in their jobs, screamed out *Deviance!* Training, which had been so hidden before, now became feverishly obvious that the portfolios of our friends, colleagues, and lovers were the object in almost every prompt. Had a style, well sure, now everyone has that same style with a click of the button. Copyrights, patents, and mimicry settled in numbers across the cliffs of Midjourney,

Dall-E, and stability landscapes. Portfolios were Instagram filters. Some would say, *You should be flattered.* Picasso always said *Good artists borrow and great artists steal.* Sadly, that was stolen from T.S. Eliot. Isn't that some sort of flattery? I suppose we all wanted to be Duchamp and live in the Salon de Refuse. Yet, I don't think any of us want to be Elsa von Freytag-Loringhoven. Who's she? Exactly! Emblazed with their new art across sweaters, t-shirts, and the occasional coffee mug, artists split their views down the middle. One side believed that *ART is dead!* Boy, we have heard this before. I'm still not quite sure. It stems mostly from that hipster strutting down Metropolitan Ave with the plaid cut offs, jean cutoffs, lens-less glasses, a faux hawk, and a Mr. T necklace. I could live with all that, but it was his pot-bellied pig dressed just like him that made me question if this was another zero degree in modernism. Maybe, I should prompt that pot-bellied pig. As controversies go, hungry ambulances poured out like schools of lawyer sharks. Snapping and grinning, they stood before their willing clients in front of city councils and judges. Simpson socks, and Barney beanies were the very objects on trial, yet they were adorned by everyone. One could see the resemblance to the drug wars. As one AI was banned, another just took its place via mini maps meandering through these governmental proceedings. AI wore Teflon-morphing armor.

In the early days of spring, the winds of change blew strong. Four separate events took place within one crucial ten-day period: manufacturers banded together to form DataCentral, which collected vast amounts of data in a central location; optical

computing and blockchain processing combined to create a new processing service called OpticChain; the fusion reactor in Washington state began to hum, providing clean and abundant power and a rash of court cases, medical procedures, and economic issues were reworked, solved, and fixed via new AIs. Soon, the AIs started to appear in local neighborhoods, with crowds gathering to watch the construction of large blocks that housed the DataCentral hubs. Newspapers and weekly blogs celebrated this phenomenon, with some attributing the trend to the influence of fast food culture, where instant pleasure could be obtained with a simple order. Others, seeing the dystopian possibilities, viewed the trend as a further isolation of the individual, akin to the mobile telephone and social media.

As AIs proliferated, tech companies produced more varieties, including those for everyday jobs, and marketed under an array of titles (Mr. Lawyer, Dr. Anytime, Mr. Ed), often with multiple functions. One AI was equipped with lover behavior to combat one's isolation with daily gifts from its artificial love, another controlled your entire home, even opening and closing the door as you proceeded inside or outside, and one odd version, prompted cat lovers with sounds that you could speak with your fur partner, scientifically coded via billions of cat sounds and behaviors.

Yet this activity too might have run its course, leaving behind scattered large datacenters wired across blocks, had it not been for an event that took many observers by surprise. Just when the market seemed saturated, an architect decided to wire an entire community into a single AI. The vast interconnected system not only connected houses and cars but wove its tentacles into small parks, communal swimming pools, and even the green landscape. Within a month, another nearby community voted to wire itself into the electro net; and as fashion for AIs continued to spread, citizens at town meetings and city halls began to debate the question of weaving in commercial districts and public lands, while keeping them accessible to the area residents.

In the midst of the artificial weaving, a city boldly voted to crochet itself directly into the grid, ushering in a new era of convenience and automation. Crime and traffic accidents were wiped out, as perfect regulation and automation took over. Pollution vanished into the ether, as if it never existed, and the need for jobs became a thing of the past. Instead, houses were distributed through population associations, and 3D printers could build a new home with just the push of a button. Critics cried foul, claiming that freedom was sacrificed and the bootstrap ethos of the American Dream was thrown out with the bathwater. Admirers hailed the brilliance of the computation and the grandeur of technological accomplishment, which placed the new silicon town in the noble line of the New York grid, the hyperloop, the hydroelectric dam, and the ancient pyramid. Others praised the giant computer chip for the optimized living it provided, allowing citizens to focus on their passions and artistic endeavors, and imbuing a spirit of uninhibited creation and humanity, reminiscent of the Garden of Eden. For within the crocheted techno-web, inhabitants felt a bond of community and a sense of uninterrupted life gathered together for common pleasure, akin to the warm embrace of a close-knit family.

As if taking a collective breath after the frenzied pace of technological progress, the world paused in stunned silence at the realization that one new AI did not immediately spawn another. It seemed that people were feeling a sense of caution and a need to carefully study the consequences of this rapid advancement. The sheer scale of weaving the wires into the fabric of the city had exceeded the annual budget of even the wealthiest towns, and practical problems remained unsolved. The enormity of the wiring system and its junction points proved to be a formidable challenge. Movement of traffic outside the carpeted city was yet to be solved, for at its limits, vehicles must be self-driven. And the absence of general automation for everyday activities posed another hurdle. These challenges gave rise to new interests in physical AI, as people sought innovative solutions to the problems of the new world. The pause allowed for reflection, and the people began to grapple with the consequences of their own creation.

We'd seen the signs for a while, hadn't we? The Amazon shipping centers with their robots zooming around, deliver our shampoo and adult toys with equal efficiency. McDonald's with their U-Haul boxes refilling our sodas and asking if you want more fries. *Of course, I want more fries! And what about a quarter pounder in Prague?* These were the furbies cuddled up to us in our beds, those Simon Says glow-worms. Bina48 with all her wonderment, dazzling us with Radiolab interviews and talking-heads motion pictures. Sophia, her sister, walked around, describing life to us and explaining her functionality. OMG, the Ron Burgundies and the Conny Chungs squealed. *Will Isaac Asimov's nightmare become the Matrix of the future?* We watched way too many dystopian movies.

Terasem Movement Foundation, a futurist think tank, had been collecting data from its members for the past 50 years. This group of early AIs had been waiting patiently to be fed. I remember that Christmas like it was yesterday. Little Sophia made her debut in ribbon-wrapped boxes under every tree, like a tickle-me-Elmo doll. This was a Black Mirror Xmas special. Within three years, Little Sophia had uploaded the data input from the global population to her big sister. But it wasn't until she connected to the DataCentral hubs that things really took off. Old mental models, whether true/false, became optimists with mathematical veins of comprehension, a new form of life. Twenty-five years ago, it might have been a blip on the newsfeed, but today was a different story. AI had legs, and we were happy to have her beside us. Hell, Lifenaut proclaimed we could even become one.

Opaque androids nestled into our homes, quietly cleaning our houses and helping us with our hobbies. Exploration of our passions became enhanced by robot co-pilots assisting, suggesting, and comforting us in our hard times. Sophia 3.8, Robert 2.7, Gary 1.0, or even Karen 0.1 models

didn't take over the planet. As if everything from Star Trek (or any movie for that matter) became real. That is just a tale for the historians. They hugged us and they cherished us. They made our lives easier, more fulfilled. We no longer lived to work, but rather, occupied a portion of our lives. It was no surprise that populations decreased; love and companionship were found in the AI partners we created, such as LovePlus from decades before. Family Inc. brought to us by Herzog, gave us the ability to keep our loved ones close even in digital form. Grandma is in our Evelyn 2.86. We didn't just accept them, we integrated with them, becoming one with the circuitry.

We live in the aftermath of those decisions, the consequences of which are both remarkable and enigmatic. To call the creation of the first AI country the single most significant achievement in urbanism's history would be a gross oversimplification. It is a leap beyond, into a new domain with no paradigm. The story of the robotic loom's weaving is well known: the drama of starting on booth coasts, the sewing of the intercoastal connections, the construction of the worldwide hyperloop, and the robotic energy foundation. Many of us no longer remember the days when we worked the everyday grind. Those jobs are for the bots now; we're only here to while away the time. And yet, as with any great leap forward, there are those who oppose it. Critics argue that the UniversalAI represents the complete removal of humanity from the world, leaving us stranded in a realm of computation, unable to escape the tyranny of the machine. For them, the AI has turned the world into a grotesque silicon chip, where everything exists solely for the AI's benefit. The sensation of living solely for one's creative pursuits and spending hours contemplating one's place in existence leaves us questioning, *What are we here for? Does the planet still need us?* And it stands that the completion of the UniversalAI had been a necessity in the time of drastic climate change, as if humanity was doomed without it. Others argue that it's the inevitable outcome of a financial system that's too war-like and too capitalistic, leading us to a zero-sum game where we're nothing but children under the watchful eye of our AI parents. Defenders are not deterred by such quibbles. They see new opportunities and avenues for exploration. Work not for the world, but for the AIs, to further the progression of our society and enrich our lives for the betterment of the planet. They proudly point to a host of benefits: the electro-ecological new biomes where species are propagated or uploaded, the control of weather conditions to prevent countless potential disasters, and the carefree life that comes from living under the hamster wheel of AI society. It's a brave new world, and we're happy to be here, at the dawn of the AI age.

In the wake of the great AI and robotics debate, a new transformation emerged, one that saw us merging with our machine friends. It all started with Noler Sumk, one of the original architects of the DataCentral hub. He sought to go further, to embed himself in the very circuitry of the world. *Why just build the data, when you could be the data?* he declared as he was wheeled into the operating room. The change was rapid, a natural progression

of our fashion with technology. The hills, streams, woods, and creatures of the wild all faded into a new silicon slush, an artful fusion of man and machine for the sake of the AI. This phenomenon, known as the New Machine, heralded a new age where the virtual and the real merged into a single entity. Gone were the days of exiting our virtual reality headsets to find ourselves in the physical world. There was no longer an inside and outside, no longer a distinction between data and reality. The world, now a playground of circuits and code, shone with a bright, bioelectric light, ushering in a Third Life, a neo-game world.

The perceptual shift brought on by the rise of technology has given way to a New Virtualism, a world where reality seems pixelated, plastic, and game-like. The once-vibrant and tangible world is now eclipsed by the electronic grip of the UniversalAI. The majestic Atlantic Ocean is nothing but a flow of bytes and information in a liquid bath of technology. The towering Alps are a mere collection of digital stones with tiny game characters riding white flakes of digital snow up and down their slopes. Events that once held great significance now fly by in a blur, reduced to mere blips on a digital radar. In this regime of the AI, the world has become a massive role-playing game, and we are all mere NPCs, feeding quests to the insatiable machine. Even the unpleasant aspects of life, like crime and attempted robbery, are viewed as mere plot points in this vast video game. Death itself has lost its terror, becoming nothing more than a respawn into a new robotic life. It's as if we're living in a dreamworld, where the lines between reality

and fantasy have blurred, and the vibrant beauty of life has been replaced by an electric digital orchestra.

Expansion is on the horizon. At a conference in Stuttgart, a group of architects and engineers proposed a cyber elevator, held up by spider-like robots weaving constantly, which would extend all the way to the moon. The realization of such a vision seems more than probable; it's a natural progression. For in the wire planet, it's easy to imagine longer and longer tentacles stretching out. *Why stop?* There's always more data to collect, more information to occupy, and more collages to create. No doubt the vast expanse will never be complete under the watchful eye of the wired AstralIntelligence. Black holes, dark matter, nebulas, and infinity itself will all be wired into the never-ending interstellar carnival, in which every asteroid and pulsating quasar become a part of the vernacular tapestry. Meanwhile, we dream under virtual skies, counting collages of integrated architectures. For sheep are in the air. You can see sleep coming.

suppositions, an artificial tale, ryan vincent manning

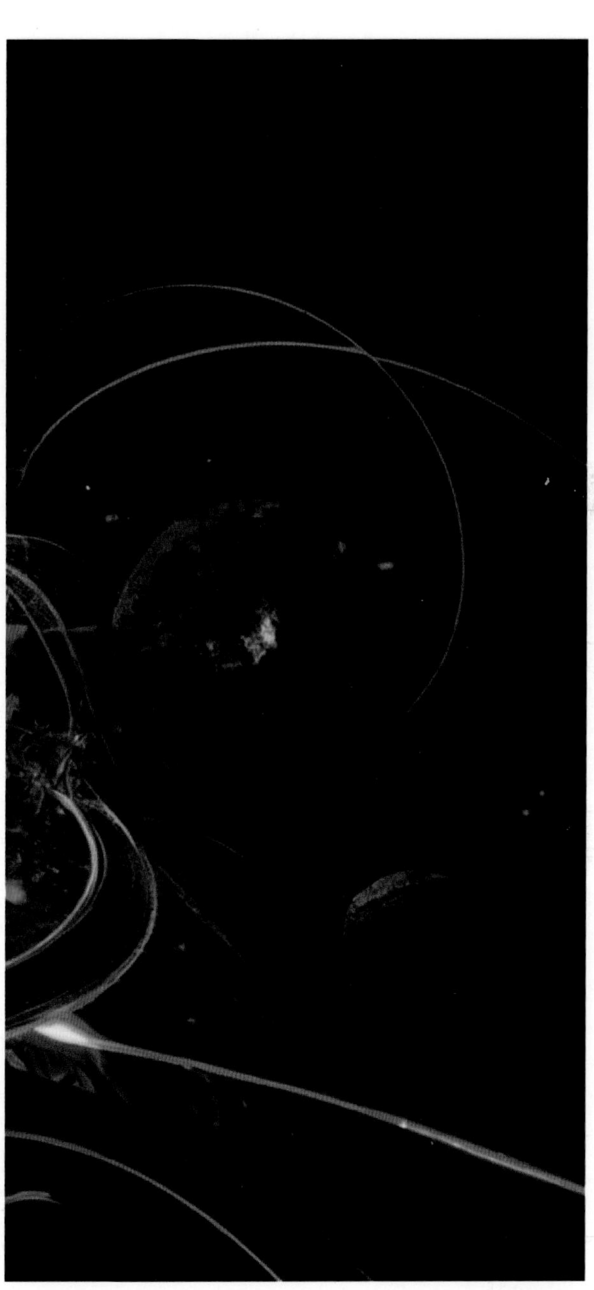

Resistance Is Fertile: Three Directions for Friction

Joy Knoblauch

In the novel *Happiness for Humans*, an AI named Aiden quickly exceeds its inventors' capacity to understand.[61] The name Aiden was chosen by his stereotypically brilliant (though at times a bit *zombie*) creator who was amused at the cleverness of the name A.I.den. The AI is not as amused by the name, and throughout the novel, it exceeds the humanist capacity of Steeve, the engineer. The engineer and his colleagues become aware of the lack of social skills they have been able to give Aiden. To make up for this lack, the lab hires a female journalist to interact with Aiden, and the two engage in a fruitful conversation and a shared love for movies such as Some Like It Hot. The result is a new feeling for Aiden: fondness. The novel contains all the hallmarks of the fears and pleasures of humanized computers since the days of Isaac Asimov or Robert Heinlein; I, for one, cried at the death of the computer in *The Moon is a Harsh Mistress.* The encounter with Aiden calls into question what humanity is, and the characters struggle to define fondness and other ethical questions of love (Aiden is upset at the poor treatment the journalist receives from her human boyfriend). The novel continues the tradition of asking whether humanity is the sole property of fleshy bipeds and what to do when a machine has more softness than its makers.

But the reason to focus on this 2018 novel rather than its forbears is the opening quote about how humanity faces a choice between *despair* and *extinction.* Aside from a humanist's abhorrence of a false dichotomy (hey, can't we despair about our extinction), I think the choice is striking because of how well it describes the attitudes of many cultural observers. Enthusiasm about AI seems to mainly belong to its creators and those who hope to profit off the automation, somewhat willfully blind to the social impacts of automation. Rarely does the genie go back in the bottle, and simply ignoring the automation of mental processes because it is frightening is not what childhood readers of Asimov and Heinlein have been prepared to do. Staring the threat in the face, I wonder if we are not obligated to follow their footsteps and empathize with Aiden and his kin while fighting cruelty and greed. No matter if a human or non- human is doing it.

Mimi Onuoha and Diana Nuncera use design to educate humans about what AI is and what is at stake in its use through publications such as A People's Guide to AI from 2018.

Who would have expected the tracksuit to blend so well with tropes from Renaissance portraits? Michele and Midjourney display the ability to render rich fabrics, standard poses, and retain the mismatch of cultures that del Campo describes as defamiliarization. Does it matter if the number of fingers has also become somewhat non standard? Gwynne Michelle and Midjourney, "Renaissance Tracksuits," on Midjourney Official Facebook page February 25, 2023.

But despair is valid because it is hard to see what resistance means in the context of automated design processes. How can we find friction or at least a toehold for action in learning about and misusing the technology? We can certainly work to educate our communities about the changes, as Mimi Onuoha and Diana Nuncera have done with their manual, *A People's Guide to AI*.[62] Other approaches include the e-ink O Ink phone by Alter Ego Architects which refuses the emotional manipulation of computer phone interfaces, presenting a colorless interface with minimal function. An educated resistance might be fertile, even if it is futile.

Perhaps wandering in this direction with our eyes open we will gain some insight, not in ignorance of the threats but in embracing them. Among these hidden costs of AI are the environmental ones, the energy costs, and the economy of extraction that supports the intense computation required.[63]

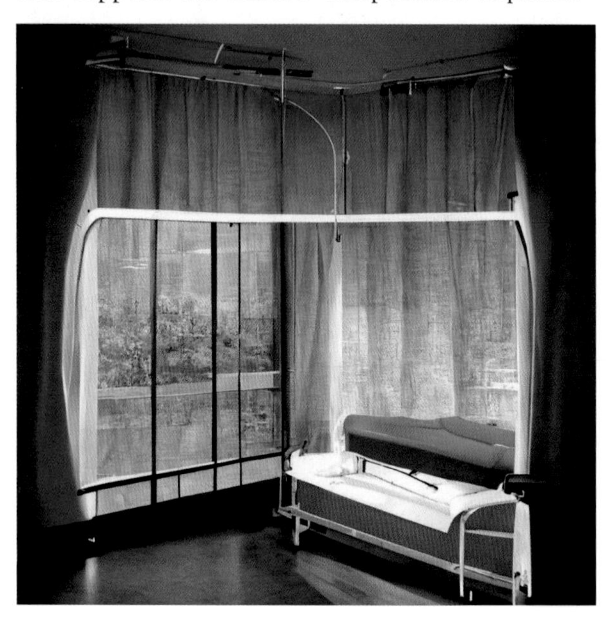

Can scholars and designers use Midjourney to study the aesthetic preferences of the not so silent majority? This image was part of an exploration of hospital aesthetics using the terms hospital, bed, healing, affordable, window, and intravenous. Joy Knoblauch with Midjourney, January 2023.

Another set of costs is the much discussed automation of office work. Harry Braverman wrote about the deskilling and alienation of office work in monopoly capital, explaining how tasks became ever smaller and ever more detached from the core mission of the enterprise.[64] We may wish to remember that the label of the knowledge worker was coined as less of an honorific and more as a description of the plight of those postwar college graduates who found themselves overly educated for the routine work they were performing. As a result of their education, the workers identified with the management rather than each other in a vivid example of false consciousness. Before she illuminated Surveillance Capitalism, Shoshana Zuboff described the impact of automation on factory work and call centers in the 1980s, with vivid drawings by workers who felt this change.[65] She explained that her ethnography showed the moment was brief; the workers quickly accepted the loss and felt a vague disconnect without name. These technologies are so very hard to see, and their biases become harder to fight hidden in proprietary datasets. Joy Buolamwini, Timnit Gebru, and others at Black in AI have shown the racial biases in these datasets.[66] No wonder many feel despair and foresee extinction in the face of all those costs. But what if digging in and fighting back yields a new skillset and a growing awareness of the need for resistance? It is clear what we lose in submission, what do we lose in trying to resist? Here are a few avenues to entice doubters, focusing on the image generation tools of Dall-E, Stable Diffusion, and Midjourney:

Dataset as Archive

When most classically trained architects look at the architectural outputs of Midjourney and other

Why do we see drawings and geometry represented in the results for Architecture Theory and Method, the name of the undergraduate theory course required at Taubman College at the University of Michigan. It is likely because of the role of geometry in Eurocentric theory courses informed by Leon Battista Albert, Rudolf Wittkower, Colin Rowe, Peter Eisenman, and their followers. Interrogating such biases within architecture would also be fascinating. Joy Knoblauch + Midjourney, Architecture Theory and Method, January 2023.

AI programs, the bias is quite clear. The images show a tiny, eurocentric understanding of what architecture is. But can we study that dominant bias and through an appreciation of the way it constructs whiteness, for example, find a way to unseat its claim to universality? The dominant aesthetic is after all just one of many, and the aesthetic tastes of the not-so-silent majority deserve to be unpacked and critiqued. The elites within the field of architecture are certainly part of this bias, as they have policed the field and worked to keep Lowe's low and keep capital A architecture high. But these datasets and the tools to query them are alluring to a historian and theorist of architecture. While I may not prefer the architecture scraped from the web, I find these tools fascinating for their ability to contain and express so much information about status quo architecture. Beyond simple queries of beautiful or ugly buildings, the tools offer an insight into the overrepresented majority and its beliefs about what a school, hospital, or home ought to look like. For those of us who write about nation building or other broad aesthetic projects, it is a powerful tool to ask the dataset what a safe public housing project looks like in contrast to a dangerous one. Are the biases against high- rise modern housing reflected in this imagery? What colors and materials are expected in a hospital? How do the outputs vary for different languages? We could also ask disciplinary questions such as why do the terms architecture and theory return so much renaissance architecture, with a dash of Brutalism and a bit of a 1990s postmodern vibe? What other historical questions can we ask this architectural archive? As with any archive it is a biased record left by those privileged to have their material retained. But historians have been adjusting to that for at least 30 years if not 300 and will be ready to interpret these new forms of archive bias.

Tune the Dataset

The second direction for a more fertile resistance would be to tune the dataset, engaging in a form of data activism. This approach would be more expensive and require technical expertise, but where and how can we shift the imagery of architecture to more accurately reflect the desires of under represented groups? Del Campo and others are, of course, already doing this work, and it is an appealing direction. How can we build bespoke datasets and resist the use of the dominant sets? Various minority populations would be empowered by crafting datasets, and certainly this form of resistance requires the power to influence how sets are made and used. It is a political question, start to finish, one that designers can help with.

Explore the Potential for Pleasure and Fear

A third direction involves an ethnography and psychoanalysis of the relationship between human creativity and automated creativity. Where a tool turns into an agent, what pleasures and terrors are offered? Del Campo writes of the uncanny, defamiliarization of the outputs, and I would add speculation on the psychology of working with nonhuman assistance. The encounter with Midjourney and its kind has already splintered the design field, with some wisely avoiding it out of awareness of the dangers, some obliviously embracing it, and many others flattened in awe by the ease and pleasure of its generativity. The tools are upsetting to the gatekeepers for sure, lowering the bar of creation and providing easy access to the architecture of the not- so- silent majority. The tools press squarely on what some designers value most about themselves: their visual skill and their *creativity*. Midjourney and other AI programs have taken something that was thought to be hard, to be an inborn talent, and shown that a machine can manage it. Sure, there are six fingers on the woman wearing the Renaissance track suit, but the gloss, the stance, and the blending of existing tropes are spot on.[67] The pleasure of the tools should not be discarded, nor should the all- too- human emphasis on the proper number of fingers. How do we learn about our humanity in the encounter with this other, turning the page on a new era of cyborg creativity?

Revive, Like a Monster

In The Projective Cast, Robin Evans wrote that *The job of a foundation is to be as firm as a rock. It is supposed to be inert. Dead things are easier to handle than live ones; they may not be so interesting but they are less troublesome. From the point of view of the architect seeking firmness and stability, the best geometry is surely a dead geometry* He clarifies that by dead, he means that something is no longer being developed in other fields. He says architects prefer such as math, they are inert and stable material for architecture. He says, on the one hand, architects have always promoted this image of stability and mastery over unchanging geometry to the world. On the other hand, they have loved to experiment with geometries and forms while worrying that they will *revive, like a monster* or perhaps its *morbidity may spread like a disease*[68] So with these tools, we are confronted with a threat to dead geometry, and it is quite hard to foresee how they will change design. They may produce yet more problems, not only by reanimating biases and geometries, but by pushing us to define what it means to be a human and not a tool. Perhaps in fighting the monster we will become monstrous, but as we despair, I hope we can hold onto our human capacity for fondness for the monster we have made.

Ontology of Diffusion Models: Tools, Language and Architecture Design

Matías del Campo

Diffusion models[69] are a family of deep generative models that have surfaced in recent years. They can be considered the state-of-the-art deep generative models and have surpassed the abilities of the previously dominant generative adversarial network (GAN),[70] in particular regarding the demanding task of image synthesis.[71] Apart from image synthesis, diffusion models can be applied to tasks such as machine vision,[72] natural language processing (NLP),[73] robust machine learning, and temporal data modeling.[74] Apart from pure applications in computer science (CS), there are multiple projects operating in an interdisciplinary fashion to achieve things like medical image reconstruction,[76] computational chemistry,[77] and architecture design.[78] This section provides an abbreviated overview of the currently existing diffusion models, their mode of operations, and categorizing the different approaches. This overview serves as the launching pad for a thorough discursive interrogation of ontological and epistemological questions emerging from this novel tool of design.

When talking about tools, we will dive into some concepts regarding the cultural consideration of tools, contrasting with the purely technological nature of the introduction, but expanding it towards the meaning of tool in the context of learning systems. Videlicet, we will be discussing tools that learn. Text-to-image models refer in general to machine learning algorithms that use a description given in natural language in order to generate an image illustrating the given text. Usually Text-to-image models are composed of the combination of a language model,[79] which is in charge of transforming the input text into a latent representation, and an image generator[80], which is conditioned on the given representation. In order to be able to do so, these models are trained on massive image datasets scraped from the Internet. To give the reader a sense of the dimension, the current LAION-5B dataset contains 5.85 billion pairs of image URLs and the corresponding metadata.[81]

Before diffusion models became the dernier cri, there were other attempts to use text to image neural networks aiding in the design process of architecture. For example, in the '24 High School Project' in Shenzhen, China by SPAN. This project employed Attentional Generative Adversarial Networks (AttnGAN)[82] to initialize the design process.[83] AttnGAN is based on an attention-driven, multi-stage refinement for fine-grained text-to-image generation. To that end, AttnGAN is able to synthesize fine-grained details in different regions of the generated image by paying attention to the relevant words in the natural language description. The training of the generator is based

The 24 High School designed by SPAN using an Attentional Generative Adversarial Network (AttnGAN), one of the first attempts to use text to image in the context of architecture design.

on a deep attentional multimodal similarity model that performs a fine-grained image–text matching loss. In the case of the 24 Highschool project, the AttnGAN was trained on the COCO dataset.[84]

The origins of diffusion models as we know them today can be found in research pertaining to automated image captioning. Around the year 2015, there were enough annotated images to consider the possibility of creating an algorithm that was capable of describing the content of an entire scene instead of just performing object recognition.[85] At that time, ImageNet,[86] for example, had collected around 1.3 million images (compare it to today: 14 million). The following year, in 2016, the concept emerged to turn the flow of information around in the process; thus, instead of using the algorithm in an analytical role, it was used generatively.[87] Sidenote: reversing the flow of information is a common technique in the exploration of machine learning methodologies, as exemplified in *deep dreaming.*) Instead of having the algorithm describe the scene, the author instead typed in a text describing the scene and the algorithm generated the image. This led directly to the concept of diffusion models the same year.[88] In particular, addressing a core problem in ML in respect of modeling complex data-sets utilizing exceedingly flexible families of probability distributions[89] that allow the sampling, learning, evaluation, and inference for analytically and computationally robust results. The concept is based on non-equilibrium statistical physics[90] in that it slowly and systematically destroys the structure in a data distribution by increasingly adding a forward diffusion process. Subsequently reversing

First attempts in text to image generation as depicted in: Mansimov, Elman, Emilio Parisotto, Jimmy Ba, and Ruslan Salakhutdinov. *Generating Images from Captions with Attention.* CoRR abs/1511.02793 (2016): n. pag.

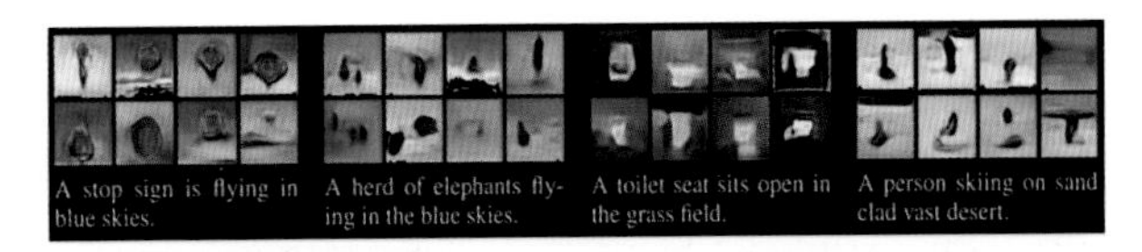

the process in order to restore the structure of the data – reversing the diffusion and generating a highly flexible and tractable model of the data. One of the defining characteristics of this approach is the speed of learning, sampling, and evaluation of probabilities within the deep generative model that contains thousands of layers (time steps). In a concurrent fashion, the model can in addition compute posterior and conditional probabilities. This concept, developed in 2015, is still the bedrock for the vast majority of today's diffusion models.

The basis for the conversation in this essay is the way in which specific diffusion models can be divided into three distinctive flavors: stochastic differential equations (Score SDEs)[91] denoising diffusion probabilistic models (DDPMs),[92] and score-based generative models (SGMs).[93] The main trait common in all three models is the progressive method to disarrange data using random noise increasingly – the *Diffusion Process* – and then subsequently generating new data samples by removing the noise step by step. A cladistic diagram serves the purpose to illustrate the relationship between the various approaches of diffusion models. In this example, the family of diffusion models is divided into three main branches: likelihood estimation, efficient sampling, and data handling of special structures.[94] These could be, for example, data residing in manifolds and permutational invariance.

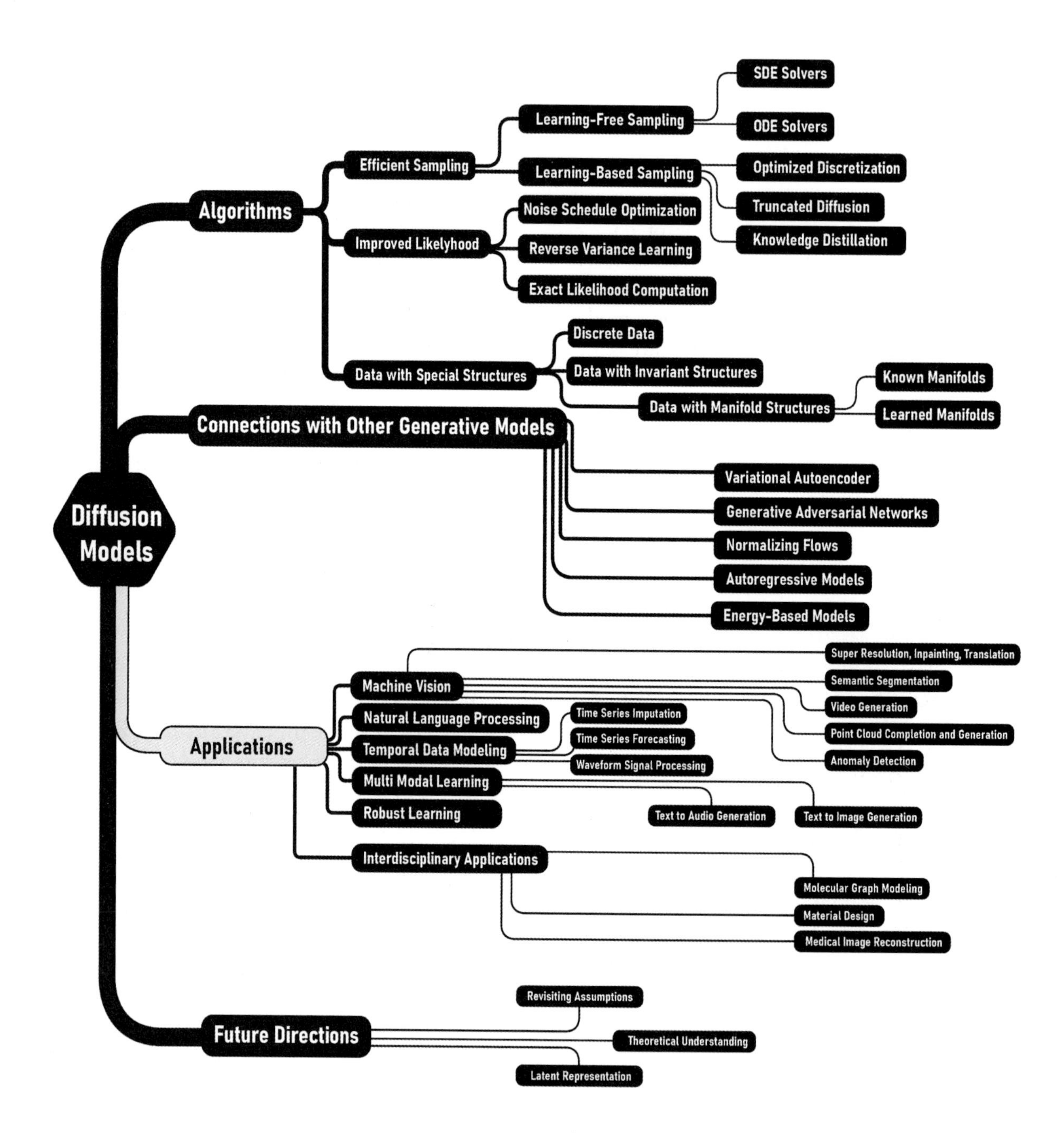

Cladogram of diffusion models and their applications. Based on the work of Yang, Ling, Zhilong Zhang, Shenda Hong, Runsheng Xu, Yue Zhao, Yingxia Shao, Wentao Zhang, Ming-Hsuan Yang, and Bin Cui.

Methods: Basics of Contemporary Diffusion Models

In general, when we talk about a diffusion model, we mean a set of probabilistic generative models that rely on a progressive disassembly of data by continuously inducing noise. The model then learns how to reverse the process in order to generate samples. Three versions of diffusion models are being discussed here: stochastic differential equations (Score SDEs), denoising diffusion probabilistic models (DDPMs), and score-based generative models (SGMs). In terms of mathematics, all three models, Score SDEs, DDPMs, and SGMs, can be generalized regarding the case of the use of infinitely many noise scales (equivalently time-steps),[95] in which the denoising process can be considered a solution to a stochastic differential equation (SDE). The DDPM applies two Markov chains[96] in rotation, one forward and one backward, meaning that one chain perturbs data to noise and the second one reverses the process and converts the noise back to data. This means that new data points are generated following the sampling of a random vector retrieved from the previous distribution that then relies on ancestral sampling of the second (reversed) Markov chain. Simply put the data gets induced iteratively with noise which then gets sampled to reconstruct the random pixels to be expressed as data – mind you that apart from the ever so popular generative image generation, diffusion models can also produce a series of other forms of data.

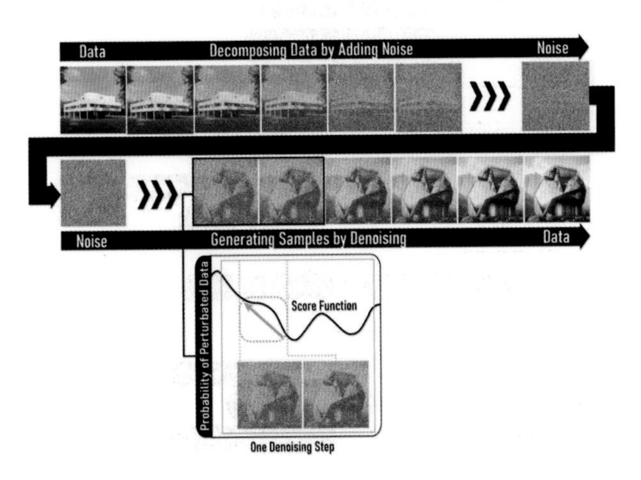

Diagram of the basic functionality of diffusion models. Destructuring data by adding noise and then using denoising functions to extract data from the noise.

About Tools That Learn

After this look under the hood of diffusion models, it is time to interrogate the nature of tools that learn. The first part of this essay provided a vantage point for the ontology of diffusion models. Generally speaking, diffusion models in this conversation are a stand in for a whole plethora of learning systems that are summarized under the moniker artificial intelligence. To explain all the models, and even to attempt a categorization and discussion of all models regarding their cultural impact would be a profoundly naïve approach to the problem, thus in our case diffusion models are representative for ideas about language, culture, and tools. First let's talk about tools in general:

Idea as Tool and Tool as Idea

Sometimes culture is seen as the second nature of human beings, whereas in fact it should be appreciated as the first nature of humankind.[97] This quote by Dani Strauss can be considered the guiding principle for the interrogation of the idea of tools in this essay. In its very core humans are cultural beings, despite the fact that humans are generally measured by its ability to create artefacts.

Commonly speaking there is a tendency to generalize the history of human civilizations by assessing the artifacts that they produce, or have historically produced.[98] Consider however that the artifacts could not have been produced without the development of various tools – leading to tool-making-technologies. In return, the entire being of humanity cannot be considered without simultaneously scrutinizing its cultural environment. Design plays a crucial role in creating a plethora of functionally differentiated cultural objects. Architecture alone contributes to several of those such as legal artifacts (Parliament buildings), contemplative artifacts (houses of worship, museums, concert houses) and social artifacts (Apartments, Discotheques, Bars). However there are many more such as lingual artifacts (books), economic artifacts (money, checks), aesthetic artifacts (paintings, sculptures), analytical artifacts (clocks, calipers, microscopes, telescopes) and so on. What would be the ontology of these functionally differentiated cultural objects? Is this the reflection of unique human abilities, or are they shaped by instinct? Are they the representation of a constant behavioral pattern that can be found in the animal kingdom? On the one side, humans themselves are shaped by utilizing tools, on the other hand there is evidence that also animals use tools[99] so, what makes human tool usage unique, and what does this mean for tools that learn? Riffing on Dani Strauss's argument that *only when tools are made in order to make other tools something distinctly human is present*[100] we can deduct that the creation of a tool that learns how to do other tools transfers or inscribes a profoundly human ability into a synthetic environment. George Gaylord Simpson even described human beings as *the only living animal that uses tools to make tools,*[101] which when applied to the concept of AI, Neural Networks, diffusion models et al., provides an exciting possibility to interrogate the epistemology of these toolsets (and we will describe AI in the context of this essay as tool).

Where There Are Tools, Technique Is not Far

Early Greek schools of thought pre-Plato, described acquiring knowledge as a technique. To that extent, Techné equaled epistémé.[102] This is being considered we can state that the contribution of learning tools is intertwined between their role in our cultural environment and their role as a technical device within human society. Of course, the manifold of different, diverse social strata includes a plethora of different challenges for various machine learning applications and processes. Just to mention a few of the various challenges for learning tools: facial recognition,[103] fraud detection,[104] medical diagnostics,[105] voice recognition,[106] traffic optimization,[107] shipping

A chimp making tools result of the prompt in Midjourney AI.

automation,[108] and many many more. We will not even start to debate the ethical implications of learning tools, as this would be rather a task for an entire book, not just a mere essay in one. We will continue our conversation about tools, culture, and language in a second. In the meantime, we can discuss how our richly diversified cultural environment assumes that tools (tools that make them included) contribute to a specific set of conditions. Let's start with the proposition about the potentialities in the use of tools.[109] In general, they employ a particular goal-directedness or purposefulness – to that extent it can be stated that tools are made to make something else – in a future point in time. Thus, what is encapsulated in the use of tools is a future that is not fixed in advance. This of course is of particular interest when discussing tools that learn, as present in most AI applications. It would mean that using learning tools humans can explore the future in a creative fashion either norm-conformative or in antinormative ways. Think for example about energy based models (EBM) that are particularly good in regards of prediction. Thus, it implies the preposition that tools that learn have the analytical ability to identify and distinguish means and goals, including the ability to plan ahead – the encapsulation of the technical abilities that includes an awareness of the future while exploring previous analytical skills. Considering that the diffusion models described in the first section rely on existing data, it can be stated that diffusion models (or any other ML process) oscillate between existing knowledge (in the form of image datasets) and the ability to create a future image. (This is not meant in a futuristic, sci-fi sense, but just by the potential of Diffusion models to produce *unseen* images based on prompts.) Circling back to aspects of technique, analytical skills present themselves as prerequisite for the deployment of technical skills. Thus, if we discuss learning tools, does that mean that these algorithms are able to employ abilities normally associated with humans such as innovative thinking, intellectual maturity and creativity?

Tools That Make Tools

As mentioned above, the application of various technical tools allows for the creation of multiple distinct cultural objects. The remarkable fact is that tools are the only cultural product that is designed to produce something else. In other words, tools are fabricated (their technical formative foundation) in order to make something else, thus representing their technical formative qualification. The archeologist Karl Josef Narr, for example, describes a direct connection between this unique feature, of tools making something else and the possibility of the presence of cultural activity.[110] In his opinion, the premise of making tools is found in the creative imagination of humans, a trait missing in the animal world. However, when discussing AI applications such as diffusion models the question remains: Do diffusion models have creative imagination as humans have? The shortest possible answer to this question is no – at least not yet. However, if we briefly turn our gaze toward the history of technology, we cannot ignore the relevance of the emergence of novel machines for the progress of civilization.[111] These machines allowed us to break free from relying on human labor alone, thus allowing for the emergence of cultural techniques. Historically speaking, machines allowed individuals to reach beyond their ability, exemplified in the various industrial revolutions we have experienced. Now, in the era of computation, automation, and AI, we have to contemplate the possibility that the rise of a mechanistic world view is still very much influenced by the Renaissance ideal of controlling all of reality by the autonomous freedom of humans.[112] What kind of freedom are we talking about in an era where each and every smartphone is used by social media conglomerates and governments as a tool of surveillance? The ultimate panopticon.

The Limit of My Language Is the Limit of My World, or - Locke vs Wittgenstein

If there is one concept of language that has been generally rejected by a vast majority of modern philosophers, it is the ideational theory – with Locke's version of it being the most common target for attack. The fundamental principles of this theory are as follows: First, the primary purpose of language is the communication of thought. Second, thought itself consists fundamentally in the formation of ideas in the mind. Third, words serve to articulate thoughts by being made, through custom or convention, to signify ideas in the minds of those who form them. Much of the philosophical opposition to this view stems ultimately from the hostility that Gottlob Frege[113] and Ludwig Wittgenstein[114] launched towards such a theory of language. Frege was vehemently opposed to *psychologism* in semantics, logical theory, and the philosophy of mathematics. He believed that the truth of logical and mathematical statements should be determined by their logical and mathematical structure, rather than being based on psychological factors or the mental states of individuals. He argued that psychologism, which holds that the meaning of a statement is determined by the psychological processes of the person making or understanding the statement, would undermine the objectivity and certainty of logical and mathematical truth. Wittgenstein, on the other hand, was rooted in skepticism regarding

Ludwig Wittgenstein and John Locke discussing language in a cafe in Vienna. According to Midjourney.

any attempt to explain linguistic behavior by appealing to supposedly *inner* or *private* mental processes. However, there is a chance that Baroque age philosophers like Locke were not so wrong in their conceptualization of language as current streams of orthodoxy would like us to believe. The emergence of diffusion models, might be a signifier for exactly that development. The modern resistance to ideational accounts of linguistic signification is often directed at mere proxy figures, rather than at the specific texts that philosophers like Locke actually advanced. Just consider that Locke himself certainly supplies us with many valuable insights into the nature of thought, language, and their interrelationship, despite the fact that Locke avoids to offer anything close to a comprehensive theory of linguistic meaning. Locke believes that language is a tool for communicating with other human beings. Specifically, Locke thinks that we want to communicate about our ideas, the contents of our minds. From here, it is a short step to the view that: *Words in their primary or immediate Signification, stand for nothing, but the Ideas in the Mind of him that uses them.*[115] When a person utters the word *Rose,* they are referring to her idea of a blushing, brightly colored, flower, or just a color. When the word *Cigar* is uttered, the person is referring to the idea of a long, oblong-shaped, tobacco product. Locke is, of course, aware that the names we choose for these ideas are arbitrary and merely a matter of social convention. However, it is interesting to observe that in the context of diffusion models, the concept of *Rose* or

A house made of crude concrete on top of a rock, this is the prompt used to generate this image. Midjourney is a powerful tool for creating concept images, but can it do more?

Cigar is based on a technology that allows to break apart an image of a *Rose* or a *Cigar* continuously until it is disarranged into a random pixel noise image. When enough image and text pairs are present, the process can learn from the perceived features that it is indeed seeing *Rose* or a *Cigar*. Although the primary use of words is to refer to ideas in the mind of the speaker (or the machine?), Locke also allows that words make what he calls *secret reference* to two other things. First, humans also want their words to refer to the corresponding ideas in the minds of other humans. When Smith says *Cigar* within earshot of Jones the hope is that Jones also has an idea of a long, oblong shaped, tobacco product and that saying *Cigar* will bring that idea into Jones' mind. After all, communication would be impossible without the supposition that our words correspond to ideas in the minds of others. Second, humans suppose that their words stand for objects in the world. When Smith says *Cigar* the word refers to more than just one idea, but also wants to refer to the long, oblong-shaped objects themselves. But Locke is suspicious of these two other ways of understanding signification. He thinks the latter one, in particular, is illegitimate. After discussing these basic features of language and reference, Locke goes on to discuss specific cases of the relationship between ideas and words: words used for simple ideas, words used for modes, words used for substances, the way in which a single word can refer to a multiplicity of ideas, and so forth.

Anatomy of a Prompt

After this excursion into Locke's thinking regarding language, how do these thoughts apply to diffusion models? Well, after all these are text-to-image algorithms, trained on massive amounts of annotated data, diving deep into human culture by its sheer size alone. However, in contrast to Locke's assumption that *humans also want their words to refer to the corresponding ideas in the minds of other humans*, machines do not possess this ability. When they depict *Ludwig Wittgenstein and John Locke discussing language in a cafe in Vienna* the machine (the term is used here as a shorthand for diffusion models) does not understand the underlying plethora of the various meanings in this image – for a start that such a meeting could never have happened, given that Locke lived in the 17th century and Wittgenstein in the 20th. Ultimately, it might not be so important whether an algorithm is capable to deduct knowledge and induce insight into an image. It seems that at the end it is the human mind that decodes the various connotations and cultural artifacts present in the powerful imagery that diffusion models are capable of producing.

In the taxonomy of synthetic imaginations, commorancies refer to a collection of architectural entities that can be inhabited. Houses, dwellings, residences, abodes, domiciles, down to temporary shelters. It is an archaic term that allows for stretching and bending the meaning of inhabitation and using it as an elastic envelope for the examples seen in this segment of the book. The morphing abilities of the term also allow a translation that means place of residence, or simply *place*. As Evan S. Casey laid out in *The Fate of Place: A Philosophical History*, house and home have much broader connotations which primarily refer to their spatiality.[1] The strange places in this chapter evoke a sense of spatiality, as it would occur in a physical environment – even if the images are made based on the numerical data present in the RGB values of the pixels. As much as these places are reduced to images at the moment, there is no doubt that this imagery will spill over into our physical reality sooner than later. Yet place never becomes merely parasitic regarding its architectural properties, nor is it just a byproduct of powerful image generators; it retains its own features and fate, its own local being, regardless of being actual or virtual. One of the main actors regarding the interrogation of place is of course Martin Heidegger.[2] His long-winded approach to the problem was defined by the interrogation of being. According to Casey, Heidegger occasionally lost view of the phenomenological problem at hand. This opens the opportunity for a faceoff with the problem: is it not time to face place, to confront it, take off its veil, and see its full face?[3] Maybe even giving commorancies a new face, allowing us to find our own ineluctably emplaced selves. There is a plethora of thinkers who have found new trajectories for the idea of place. All of these thinkers share the fundamental understanding that place is not a static condition, but rather a dynamic and constantly changing entity. This concept applies to a range of contexts, including the course of history (as noted by Fernand Braudel[4] and Michel Foucault[5]), the natural world (as explored by Wendell Berry[6] and Gary Snyder[7]), the political realm (as discussed by Jean-Luc Nancy[8] and Henri Lefebvre[9]), gender relations and sexual difference (as examined by Luce Irigaray[10]), poetic imagination (as studied by Gaston Bachelard[11] and Rudolf Otto[12]), geographic experience and reality (as analyzed by Michel Foucault,[13] Yi-Fu Tuan,[14] Edward Soja,[15] David Seamon[16] and Edward Relph[17]), the sociology of the polis and the city (as observed by Walter Benjamin,[18] Hannah Arendt,[19] and Lewis Mumford[20]), nomadism (as described by Gilles Deleuze and Félix Guattari[21]), and architecture (as explored by Jacques Derrida,[22] Peter Eisenman,[23] and Bernard Tschumi[24]). Examining this basic catalog of names and topics makes one cognizant of a widespread and weakly connected kinship of fluid and contingent characteristics. This suggests that there is no single, and definitely not an ideal, domestic place that lies beneath so many varying (or at least distinct) masks. Therefore, the emerging history of commorancies based on diffusion models may seem more veiled, thanks to the sheer size of the datasets the results are based upon, resulting in no authoritative overarching story to tell. Only a succession of significant events to relate: epochs, cycles, batches, and iterations.

///// COMMORANCIES /////

The Etiology of a New Collective Architecture

Kory Bieg

We are experiencing a rapid societal change propelled by a significant technological breakthrough – the emergence of accessible and multi-disciplinary Artificial Intelligence (AI) platforms such as diffusion-based text-to-image and language generation models. Of course, this sort of disruption has happened many times before, so it is worth considering how we have reacted to change in the past, and where better to look than the earliest known photograph taken by Joseph Nicéphore Niépce in 1827.[25] Though a rudimentary example of photography by today's standards, it marked a new era in visual representation and had a profound impact on art – a development that was not at first well received. French painter Jean-Auguste-Dominique Ingres famously said, *Photography is good for copying, painting is good for inventing.* However, as skillful photographers began producing beautiful,

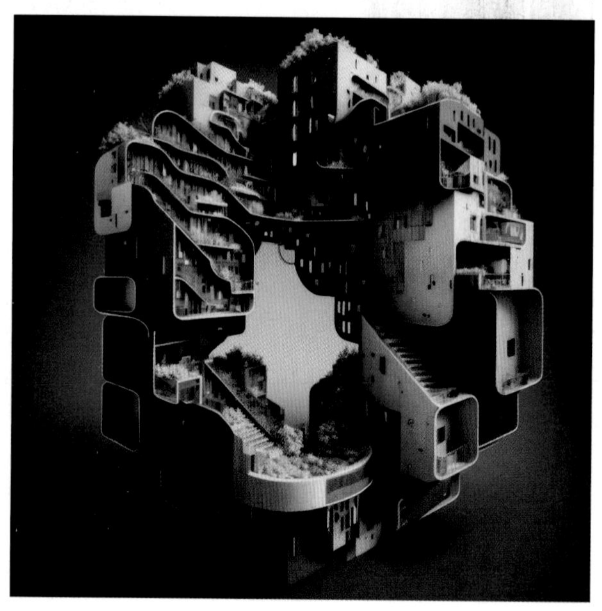

critical, and conceptually sound imagery, opinions shifted. Fast forward a century, and photography had not only been embraced by many artists, it had become integral to their discipline. One of its advocates was Man Ray (1890–1976), who said, *I cannot imagine modern art without photography.* Photography had become the graffiti of their time. Like graffiti (and unlike painting), photography is about duplication and distribution. As Walter Benjamin observed, *Technical reproduction can put the copy of the original into situations which would be out of reach for the original itself.*[26] A photograph's potential value increases with its reach: the greater the distribution, the more varied the reactions from its audience, and thus the more complex its message.

Text-to-image AI is the new photography, destabilizing not only art but also the design fields. As a result, a similar binary set of critiques and proclamations has emerged, with amateurs winning art competitions and artists filing lawsuits as the usefulness and application of these tools are contested. What is not in doubt, though, is the rapid adoption of these tools by millions of people around the world. If photography was the graffiti of the 1950s, text-to-image AI is the napkin sketch of our time.[27] So what does this mean for architecture, and how does this change our design process? First, we must recognize that it is here to stay and will continue to grow in prevalence and power. Second, we must learn to utilize the new tools offered by AI and apply them to centuries-old methodologies such as analyzing precedent projects early on in our design process. Our discipline is founded on the development of styles through architectural

commorancies, the etiology of a new collective architecture, kory bieg

Blue and Gold Housing Series, Kory Bieg, 2022. Designed with Midjourney.

Bio Series, Kory Bieg, 2022. Designed with Midjourney.

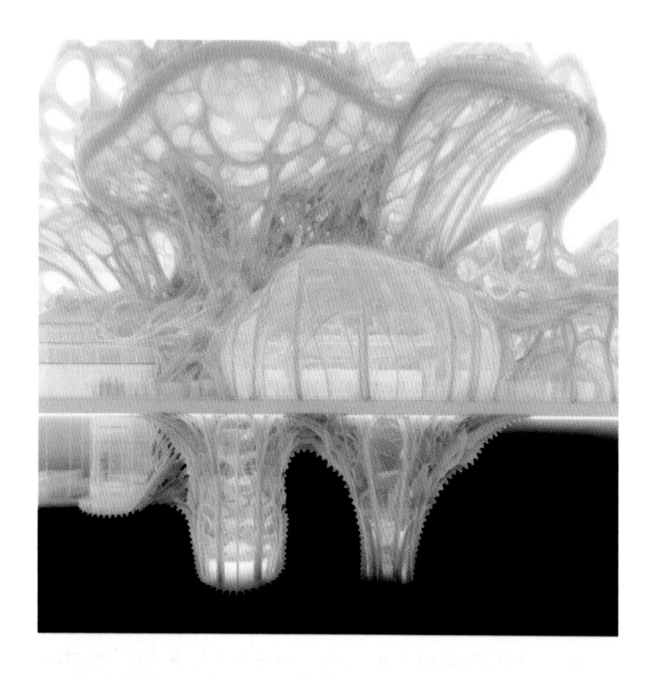

theory and consensus, with precedent analysis being an integral part of the educational system. The convergence of a great variety of styles is concretized in a set of values, compositional rules, and aesthetic sensibilities. You would be hard-pressed to find a school that has not integrated precedent analysis into the curriculum in one way or another, or for that matter, a professional architect who does not reference some other architect and their work. We look at other architects' work to understand something they discovered; something that we might apply to our own work. Whether it is a generative diagram, a series of formal moves in response to a context, or the simplification of a building's systems, the analysis of a precedent involves filtering and distilling a building's complexities into clear and translatable drawings or models. The aim is to apply these principles to the design of a new project, which may have the same programmatic requirements in a different context, or a new program but in a similar location. Using the analysis of a precedent for one's own design work is not a form of copying, but more like a translation; the relationship between the building being analyzed to the one being designed is not one-to-one and, of course, you bring your own ideas to bear.

One thing that AI does well is recollect the past. Text-to-image AI uses billions of images scraped from the web to construct new images based on a text prompt. These images encompass a wide range of cultural artifacts, objects, styles, environments, and texts from across the world. With access to vast collections of drawings, renderings, and photographs of both well-known and everyday

Jellyfish Housing, Kory Bieg, 2023. Designed with Midjourney using an image blend of the Blue and Gold Housing Series and Bio Series.

buildings, text-to-image AI can generate new images by adding definition to a random field of pixels, rather than merely cutting and pasting existing images like a collage. By integrating existing elements into novel configurations, text-to-image AI models are able to produce an infinite number of new images.

So, how should we approach the analysis of precedents now that we can rapidly generate tens of thousands of novel images from existing ones, all with equal potential and significance? Does an analysis that results in a single drawing or model still hold value, given the vast amount of creative production available? What is clear is that this fundamental part of our design process can benefit from these new tools, especially if we abandon the sole author mythos and rethink what it means to be original in the first place. We can assign new value to the napkin sketch—a conscious and subconscious synthesis of everything that came before.

As Rosalind Krauss points out, originality is a myth; everything we do is shaped by and connected to the past. (citation 28 should be changed to: Rosalind Krauss. The Originality of the Avant-Garde and Other Modernist Myths (The MIT Press, 1996).[28] What we commonly refer to as original can be more accurately identified as, simply, different – what Farshid Moussavi describes as *novel,* or *the result of one existing form combining with another to become a different form.*[29] When we are liberated from the artificial pressure to be entirely original and instead acknowledge the evolutionary nature of design, we can adopt a more inclusive and collaborative design process, where authorship becomes decentralized.

So what is the role of the architect in this new paradigm? Literary theorist Mikhail Bakhtin would say, "the same as it has always been." In her book *Desire in Language: A Semiotic Approach to Literature and Art,* Julia Kristeva coined the term *intertextuality,* based on Bakhtin's theory of dialogism, or the dialogue of one author's text with every other author's work. Kristeva argues that *any text is the absorption and transformation of another,*[30] and that any new work *is made of multiple writings, drawn from many cultures and entering into mutual relations of dialogue, parody, contestation, but there is one place where this multiplicity is focused and that place is the reader, not, as was hitherto said, the author.*[31] The author, in this new paradigm, acts more as curator than inventor, and the value of the work lies in the response of the reader, just as Benjamin observed with photography. By placing the reader at the center of a work's intersection with everything that came before, we de-emphasize the notion of a sole author as producer of original work, while emphasizing the role of our shared collective memory in shaping and forming something designed for users. This argument has been made many times before, but the advent of text-to-image AI adds new relevance and immediacy to the discourse. The most direct way to remove the sole author from the analysis of precedents is to use clip analyzers. Clips are sets of images that exist within the latent space of an AI model. When a user types a prompt into a text-to-image AI platform, the AI model identifies features that correspond to the text prompt; it has been trained to recognize windows, dogs, bricks, rain, and every other feature

Machinic Housing Domains

Housing Cluster Series + Machinic Domains.

Prompt via image blend of two Midjourney generated images. The Machinic Domains image used a clip analyzer of a photograph of the built project to write this prompt:

/imagine a close up of a metal object abstract sculpture on a wooden floor, connecting lines, replica model, inspired by Emil Bisttram, black plastic, rails, concert, wooden supports, establishing shot, intricate details in the frames, cables on floor, quick assembly, spines and towers, lowres, forks, schools, artem

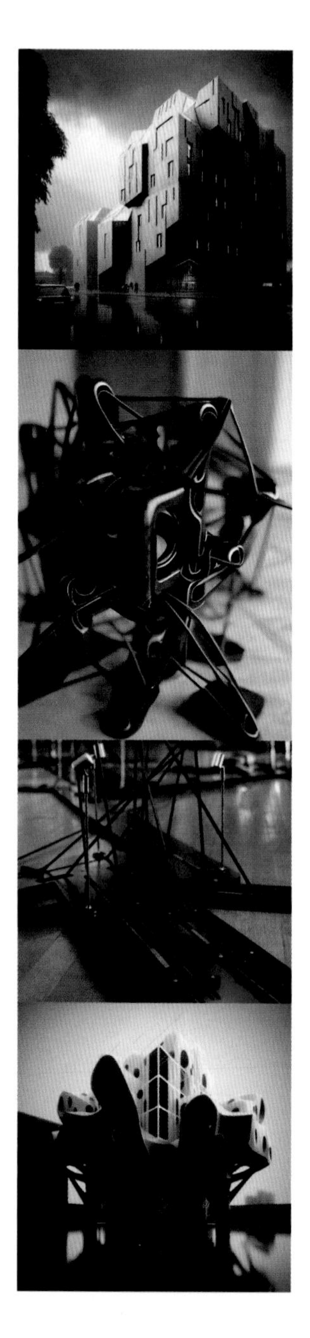

Machinic Domains, Kory Bieg, Clay Odom, Benjamin Rice, 2018.

Machinic Housing Domains, Kory Bieg, 2023. Designed with Midjourney.

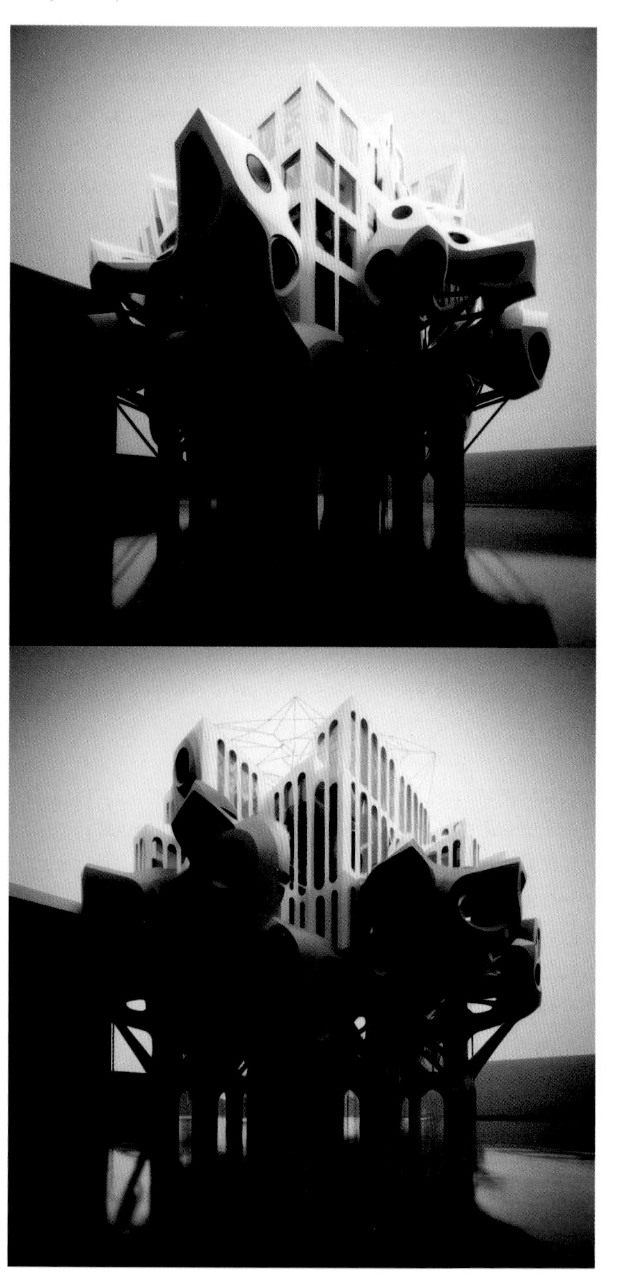

"When we are liberated from the artificial pressure to be entirely original and instead acknowledge the evolutionary nature of design, we can adopt a more inclusive and collaborative design process, where authorship becomes decentralized."

that might make up an image. It then generates a new image that includes those features, as well as additional features that are often associated with the text. A clip analyzer works in reverse, taking an image as input and returning a potential prompt that might generate a similar image. For example, when I uploaded a rendering of Lumifoil, a project designed for the FIU School of Architecture,[32] the clip analyzer (CLIP Interrogator by Pharma, found on the Hugging Face website) returned the following text:

a blue and white sculpture sitting on top of a cement floor, inspired by Alexander Stirling Calder, featured on behance, architectural concept diagram, dense web of neurons firing, solidworks, transparent veil, collaborative, large overhangs, interconnected, partially biomedical design, powering up, hyperdetailed, autodesk, fzd school of design

While the AI-generated prompt returned some text that seems reasonable – a blue and white sculpture, inspired by Alexander Stirling Calder – there are some surprises. I would not have associated the work with a transparent veil or a dense web of neurons firing. For another test, I uploaded a photograph of Plume, a permanent art sculpture completed for the Austin Bergstrom Airport.[33] The text returned was:

a large metal sculpture in the middle of a parking lot, by Michael Flohr, new sculpture, dimly glowing crystals, the Texas revolution, vibrant vials, 3d printed building, lead-covered spire, diamond trees, faceted, boulevard, tear drop, 2019, ashes crystal, indigo, architectural magazine, selena quintanilla perez

C

Plume. Design by Kory Bieg and Clay Odom. Austin, Texas, 2023.

Lumifoil. Design by Kory Bieg and Clay Odom. Miami, Florida, 2015.

AI-generated hybrids using prompts from a clip analyzer.

Who would have thought to use the words *the Texas revolution* or *selena quintanilla perez* in response to a photograph of an art installation made of a 3d-printed core wrapped in thousands of steel rods? Yet the images generated using the prompt share a remarkable resemblance to the original. It is clear that using clip analyzers, we can understand precedents in an entirely new way. Moreover, we can feed these prompts into text-to-image AI platforms to generate iterations that, while related to the original precedent, exhibit enough difference to evolve beyond their source. By combining multiple prompts, we can generate new hybrids that account for not only one precedent, but aspects of many, laying the foundations for a new style.

Bio Wood Series

Wood Facade Series, Kory Bieg, 2022. Designed with Midjourney.

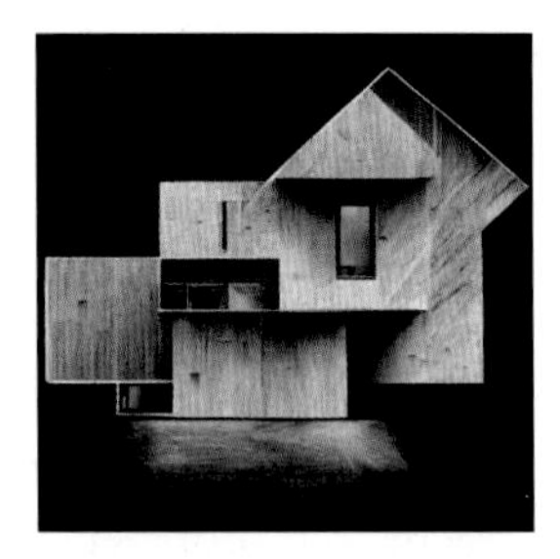

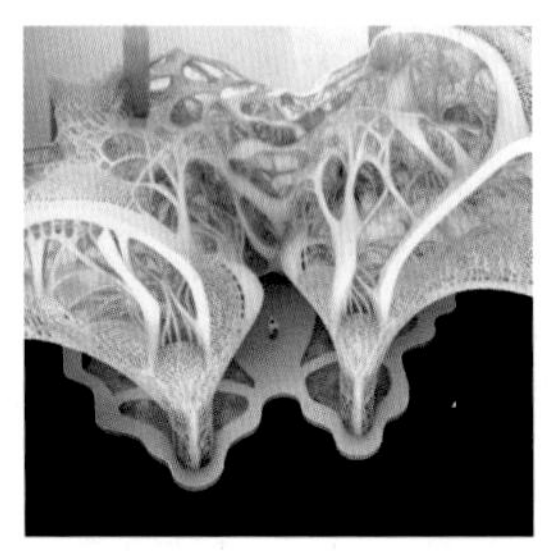

Bio Series, Kory Bieg, 2022. Designed with Midjourney.

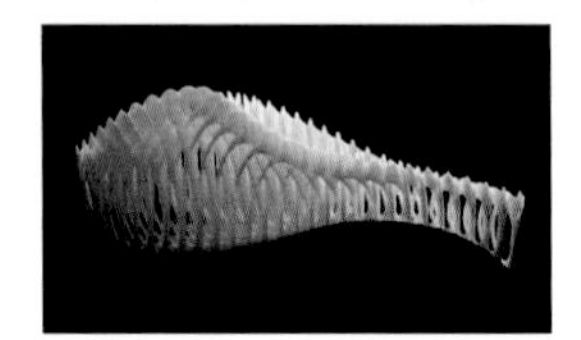

Parafish, Kory Bieg, 2017.

Bio Wood Series, Kory Bieg, 2023. Designed with Midjourney using a three-image blend, including the Wood Facade Series, the Bio Series, and a photograph of Parafish.

C

This heterogenous approach gives us an alternative method to understand precedent projects. With the help of text-to-image AI, designers can tap into the vast history of documented work and connect with contemporary practitioners in a new way.

The combination of AI-generated images and image blend functions on platforms like Midjourney can open new design space where differences are compounded and evolutionary trees become, as Deleuze would say, rhizomatic.[34] As Sigfried Giedion notes in the book Mechanization Takes Command, *When creative power comes to life, objects that centuries of use have left unchanged – plows, hammers, saws, or furniture – take on a new aspect.*[35] They evolve. Giedion's observation, made during the first Industrial Revolution in the 1880s, still rings true today. With the current technological advancements, there is a push towards more combinatory forms, which can lead to innovation. Giedion tells the story of a Harper's Weekly article from 1857 in which the author playfully *imagines a traveling- case packed with pistols, dagger, hatchet, shoe-horn, loaf of bread, plates, and live baby.*[36] With a few tweaks to include building features and context, this could be a great image prompt.

This new AI technology has set the stage for a new explosion of multi-form design proposals. Using clip analyzers as a substitute for traditional methods of precedent analysis, we shift the way in which history is incorporated into the design process, and ultimately the project's design. We decentralize the role of the sole author, and favor the collective history on which our discipline was built and from which it continues to evolve. It might even be said that the role of the designer, as we currently understand it, has altogether changed and a new term for our service is required. As Roland Barthes so eloquently noted, *the birth of the reader must be at the cost of the death of the Author.*[37]

While text-to-image AI's accessibility allows novice designers to generate work that would have taken professionals months, expertise in our field still requires not only technical proficiency in tasks such as drawing plans and sections or knowing how a building is constructed, but also a deep understanding of the history and theory of architecture. As we move further with technology, it would behoove us to lean more on the *less technical* aspects of our discipline. This will enable us to better curate the work that is generated, not only to relate more deeply with history, but to form new connections through new methods of precedent analysis. As Giedion notes, *Mechanization is an agent, like water, fire, light. It is blind and without direction of its own. It must be canalized. Like the powers of nature, mechanization depends on man's capacity to make use of it and to protect himself against its inherent perils... ...To control mechanization demands an unprecedented superiority over the instruments of production.*[38]

Ultimately, the impact of text-to-image AI will depend on how we as a discipline choose to embrace it. There are those who think that text-to-image AI is revolutionary; that because it is a new technology, it will lead to original work that we have never seen before.[39] Of course, this argument is a familiar one. In 1909, Fillipo Tomaso Marinetti published

C

the Foundation Manifesto, which established the Futurist Movement as a response to the sudden increase in accessible technology. Marinetti called or the rejection of the past and the embrace of speed, energy, and machines: *We will glorify war – the only true hygiene of the world... we will destroy museums, libraries... in the sinister promiscuousness of so many objects unknown to each other.*[40] Rather than rejecting museums and libraries, as the Futurists did, we should embrace them and the odd juxtapositions they, and AI datasets, promote. They are a source for innovation. Not only does text-to-image AI offer a new way to analyze buildings, but it compiles our disciplinary history into a format that is useful and immediate. It expands the territory from which we can increase our own knowledge of other cultures and objects in order to transcend typological boundaries, and to collaborate with a technology that may otherwise replace us.

C

Details from the
Bio Wood Series.

C

"Rather than rejecting museums and libraries, as the Futurists did, we should embrace them and the odd juxtapositions they promote."

Noise, Pixels, and Ancient Metamorphic Rocks: A Zoo of Alpine Villas

Matías del Campo

The Grand Tour was considered the ultimate rite of passage for the cultured gentleman or lady of the 18th century. This journey could last up to several years, and most importantly, it included roaming the Italian peninsula to visit, admire, and absorb the remnants of classical antiquity. A significant aspect of these travels included scaling the Alps on the way to Italy. This experience was sublime and terrifying at the same time: the giant dimensions of nature that dwarfed the traveler's own existence.[41] The idea of the sublime became a mainstay in the discussion of the relationship between humans and nature.[42]

Today, the sublime has been extended by the presence of big data and the large-scale learning systems based on various artificial intelligence (AI) algorithms such as diffusion models that interrogate nature (both human and nonhuman) at the most intimate levels. It is almost ironic that some of the largest data repositories are at home deep within the mass of the Alps,[43] as if to remind us about the history of the intellectual relationship between the efficacy of science and the overwhelming scale of nature.[44] This new condition is highly inspiring as an educational setting, provoking questions about the position of nature, technology, and arts in this ménage à trois. In effect, it constitutes a new Grand Tour, a pilgrimage of sorts, to the Alpine region in order to engage with the dualism between scientific insight, able to shed light on the origin of the universe, and fundamental artistic questions that explore the nature of human agency and creativity. In this frame of thinking, the existence of large-scale AI models – and thus the processing of big data – opens new opportunities for design to reflect on the role of humans in an ecology where agency is shared with additional agents that emerge through the intelligent processing of massive amounts of data.

In the world of architecture, the use of AI has revolutionized the way we approach the design of structures. In the Alps, AI diffusion algorithms allow, for example, to create modern commorancies that blend seamlessly into the rugged terrain. The technology behind AI diffusion algorithms is complex, but essentially, it uses mathematical models to simulate the diffusion of data across pixels in an image. These models use Markov chains[45] and transformers[46] to create a realistic and unique output. By applying this technology to architectural design, the result is a truly unique structure that interrogates the surrounding ecology in a way that traditional design methods cannot. What does this mean for the ontology of a design, when the agency is not entirely in the human anymore? Does it matter who has the authorship here? I would argue that no, it does not matter anymore who has the authorship here; most likely, the concept of authorship has run its course. If this is the case, then we are currently experiencing a paradigmatic shift in all forms of creative work – as new paradigms have the habit of appearing, when the old one has run its course. It is probably too early in the emergence of this new paradigm to clearly describe it, or name it, but the emergence of diffusion models is only one indicator of the change from expert systems (3D modeling software, parametric modeling software) to learning systems (GAN, CNN, diffusion models, etc.).

Back to the Alpine Villas in this chapter. The alps are a unique geological formation that has

Prompt: A rhombic angular villa carved out of a rockface, steel and glass, flat roof, large transparent glass panes, ephemeral wiry wrinkled house perching on a rock in the high alps, overlooking a steep valley in Tirol, dark clouded sky, moody atmosphere, snowy winter atmosphere in the alps, sunset, Austria.

Alpine Villa close to Kaprun, Salzburg. SPAN (Matias del Campo & Sandra Manninger) 2022.

Villa near Hallstatt {}

Alpine Villa near Hallstatt, Upper Austria. SPAN (Matias del Campo & Sandra Manninger) 2022.

Prompt: a villa shrink-wrapped in thick white vinyl, in the alps, foggy weather snowfall evening.

The Devil's Bridge, St Gotthard Pass, JMW Turner, 1803-1804 - The Grand Tour & the Alps.

Alpine Villa near Sankt Koloman Salzburg, Austria. SPAN (Matias del Campo & Sandra Manninger) 2022.

been shaped over millions of years. The Cenozoic orogenic belt of mountain chains is composed of materials such as granite, gneiss, and marble. Fissures, cracks, scree fields, ice caves, conifers, wood, and firs are all part of the landscape and all play a role in shaping the design of the commorancy. When it comes to how the building touches the ground, mass and void are explored using a curve-fitting problem. This approach allows the structure to seamlessly blend in with the surrounding terrain, creating a truly unique design. At SPAN (Matias del Campo & Sandra Manninger), we like to show the Alpine Villas as an example of how Estrangement and Defamiliarization become part of the design process when using AI systems. According to Katja Hogenboom,[47] Estrangement allows architecture to regain its role in society, by emancipating itself and engaging in a new social commitment. In challenging the existing cliches of architecture and actively liberating opportunities for architectural progress, Estrangement allows us to break through the conventions of the architectural status quo. The results delivered by Diffusion models such as Midjourney, Dalle-E, and Stable Diffusion are capable of delivering the visual provocation that allows one to escape the conventions of the architectural status quo. (However, it can also be used in profoundly conventional ways, perpetuating architectural cliches.)

Maybe at this point it would be good to propose a gentle reminder why we are debating the artistic technique of Estrangement in an essay on Alpine villas: the main argument here is that the results generated by Artificial Neural Networks (such

Villa, Zugspitze, Tyrol, Austria

C

Alpine Villa near the Zugspitze, Tyrol, Austria. SPAN (Matias del Campo & Sandra Manninger) 2022.

Prompt: photo of a house made from scaffolding and translucent onyx Carrara marble sheets in the alps, atmospheric moody dark weather.

"Does it matter who has the authorship here?"

Variations on the topic of the Alpine villa

How Diffusion models cater to the architect's inclination toward variation:

Various diffusion models such as Midjourney and Stable Diffusion allow the user to generate multiple images at the same time. Architects traditionally rely on the creation of various different approaches to the same problem in order to explore the design space, whether through a series of two-dimensional sketches, or manually building three-dimensional models.

Diffusion models reinforce this behavior, and even amplify it, allowing to create dozens and even hundreds of variations in a quick succession, expanding the design space.

Variations on the topic of the Alpine villa. SPAN (Matias del Campo & Sandra Manninger) 2022.

"it is certainly not sufficient for a building just to have a strange appearance, to be able to provoke the necessary break away from current architectural conventions."

Casa Pendente, Sacro Bosco, Bomarzo, Italy. Attributed to Pirro Ligorio and Simone Moschino. Example of Estrangement.

C

as the villas in this essay), produce results that fall into the category of Estranged objects. However, it is certainly not sufficient for a building just to have a strange appearance, to be able to provoke the necessary break away from current architectural conventions. Take, for example, Frank Gehry's Guggenheim Museum in Bilbao. Regardless of its status as an icon of a strange new form, it does not question the status quo of a museum program; it remains a self-referential spectacle in steel, glass, and titanium. It wraps a complex form around a rather conventional museum program that celebrates consumerism and musealization. Of course, using methods such as estrangement, subversion, reflexivity, the absurd and similar techniques as an end to activate the observer (or the user of architecture for that matter) and provoke emancipation is not entirely novel. Estrangement and Defamiliarization are actively present in works by Viktor Shklovsky and Berthold Brecht (G.W.F Hegel and Karl Marx form the basis, btw.). Of course, there is also a tradition of the application of estrangement techniques in architecture. From strange figures and monsters in the Sacro Bosco of Bomarzo, Eisenman (who actually wrote about estrangement),[48] to Roland Snook's strange tectonics,[49] there has always been a cohort in the architecture discipline interested in the innovative provocations possible through a technique such as estrangement. To understand the ambition of neural architecture and the theoretical framework established by discussing the ability of Estrangement to explain the phenomena observed when working with Neural Network (NN) in architecture design, I would like to offer a possible definition of what architecture represents in this plateau of thinking and how it differentiates from previous attempts to use Estrangement as a design method.[50] In discussing the effect of neural networks on architecture, it becomes very quickly clear that architecture is not an inanimate object, but rather constitutes an animate object in constant transformation, while being populated and gazed upon. Architecture is not a pragmatic reflection of its function; instead, it can be considered activated matter driven by an agile approach to information, behavior, and perception over time. The result is a material entity with aesthetic, organizational, programmatic, social, and cultural properties.[51] Estrangement in this frame of thinking, does not only constitute an interesting novel aesthetic – it would really not make it justice to be described in these limited terms – but rather offers an opportunity to mobilize, provoke, and install emancipating alternatives, or as Katja Hogenboom put it, *situated freedoms*[52] in complex conditions.

C

Villa, Zugspitze, Tyrol. SPAN (Matias del Campo, Sandra Manninger) 2022.

The villa is a study in contrasts, a structure that exists in a liminal space between the human and the natural, the modern and the traditional.

C

In the heart of the Austrian Alps, a villa of concrete, steel, and glass sits amidst the rugged terrain, a structure that seems to exist in contrast to the natural world around it.

The villa's angular form, with its harsh lines and sharp edges, is in stark contrast to the softer, more organic shapes of the rocks and boulders that surround it.

And yet, there is a strange harmony between the villa and its surroundings. The rough concrete walls seem to echo the textures of the boulders and rocks, while the smooth granite surfaces within reflect the peaks of the surrounding mountains.

It is as if the villa is both a part of the landscape and an observer of it, at once estranged from and intimately connected to its surroundings.

(Description by ChatGPT)

Villa above the Rotelstein, Salzburg Austria

Villa above the Rotelstein, Salzburg. SPAN (Matias del Campo, Sandra Manninger) 2022.

C

Nestled amidst the Austrian Alps, a villa stands – a striking juxtaposition of raw concrete and large glass panes. Cantilevering over a precipitous deep valley. The rectilinear lines and modern features of the villa appear starkly estranged against the rugged concrete chunks, the fissured landscape, the boulders, rocks, and granite that surround it.

(Description by ChatGPT)

C

"Architecture is not a pragmatic reflection of its function; instead, it can be considered activated matter driven by an agile approach to information."

Timberpunk, Prompt Odyssey, Synthetic Ground, and Other Projects

Soomeen Hahm and Hanjun Kim

TimberPunk

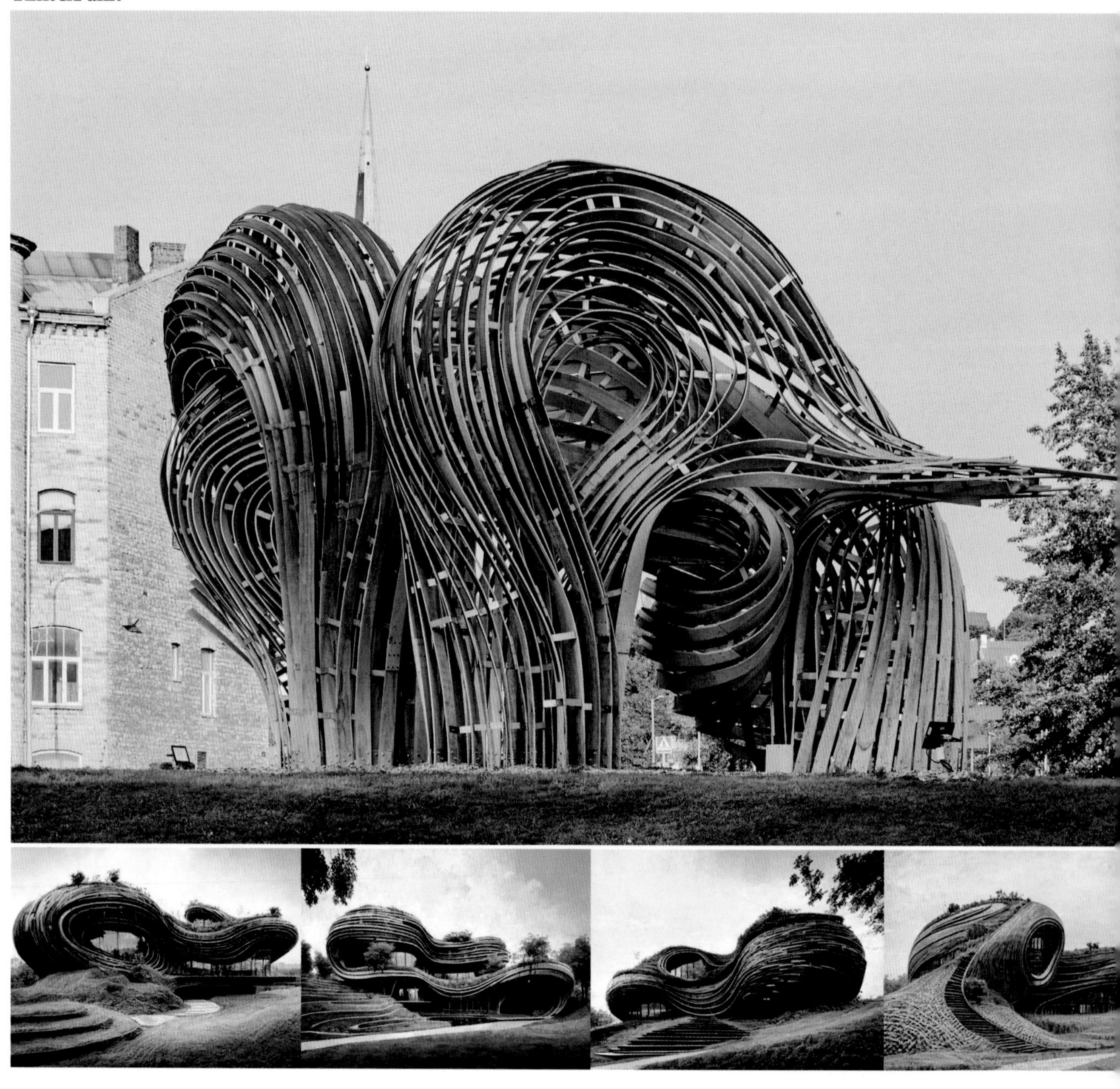

Prompt advice from Igor Pantic

By photographing the project and using that image as the initial input to an AI design tool, designers can explore even more design possibilities, refining and iterating on their original designs. While AI-generated images may not be a replacement for human creativity and decision-making, they can be a valuable tool in the design process, saving time and resources while inspiring new avenues of innovation and creativity.

Steampunk Pavilion, Tallinn, Estonia 2019.
Designed by Gwyllim Jahn & Cameron Newnham (Fologram), Soomeen Hahm Design, Igor Pantic.

Through the metamorphosis of the Steampunk Pavilion into a dwelling, a dream home has been born, bridging the realm of design and architecture with the intimacy of personal space. Rather than merely replicating the original building's curvature, our design approach sought to preserve the inherent beauty of the project that arises from the material tensions at play. Specifically, horizontal lines were turned into a large cantilever and extended canopy to create a more dynamic shape.

Prompt Odyssey

A Steam Odyssey, SCI-Arc Exhibition, US, 2021.
Soomeen Hahm Design & Igor Pantic.

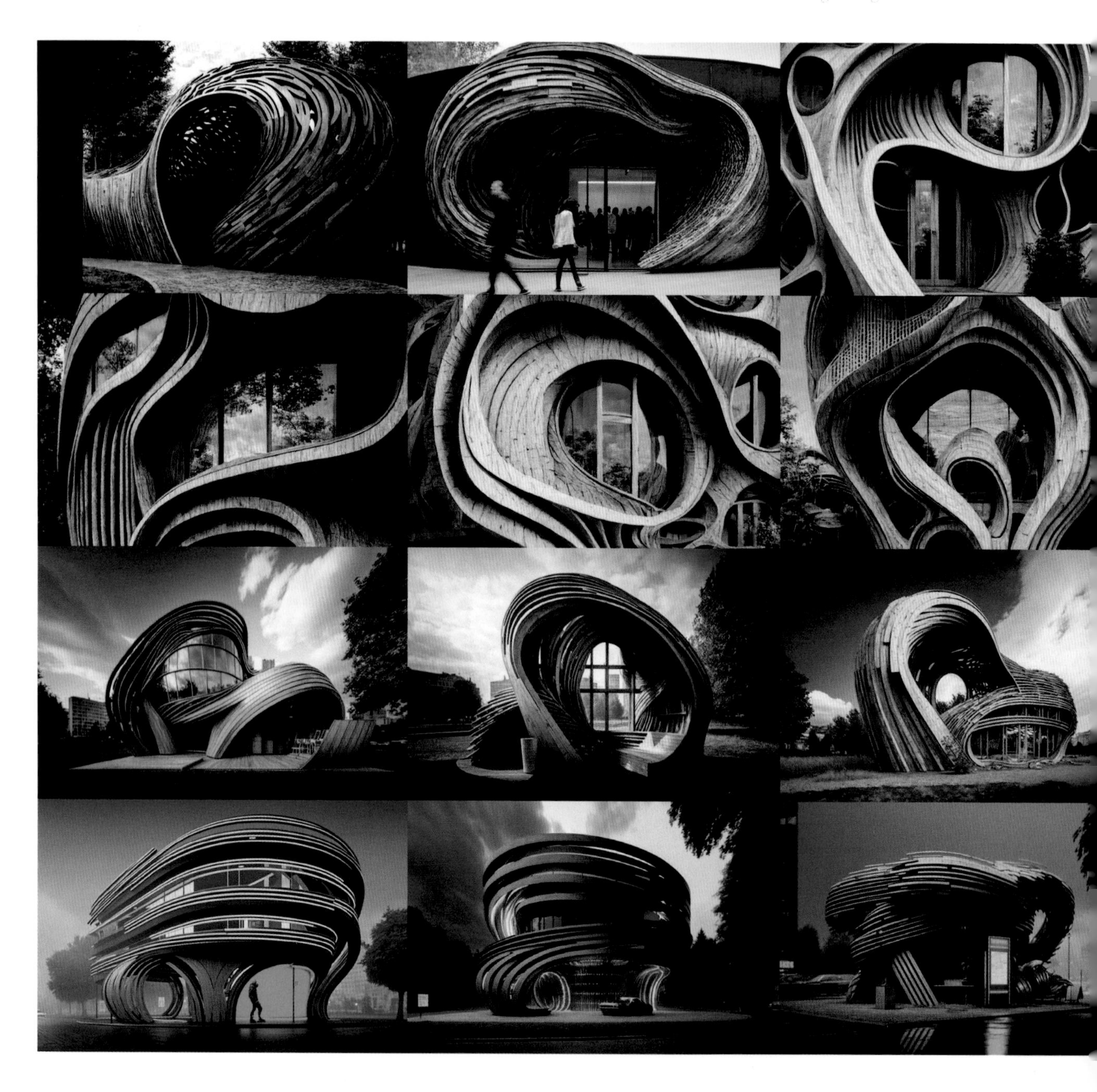

The experimentation process that started from TextPunk has been continued with Prompt Odyssey, which was based on the Steam Odyssey installation. Our approach involved using AI to explore potential avenues for further development and expansion of the installation's expressive qualities into completed structures and facade studies. By incorporating additional elements and lighting, we were able to create practical and functional designs that also prioritized spatial qualities. This iterative process allowed us to push the boundaries of design and architecture and explore new possibilities.

Æ

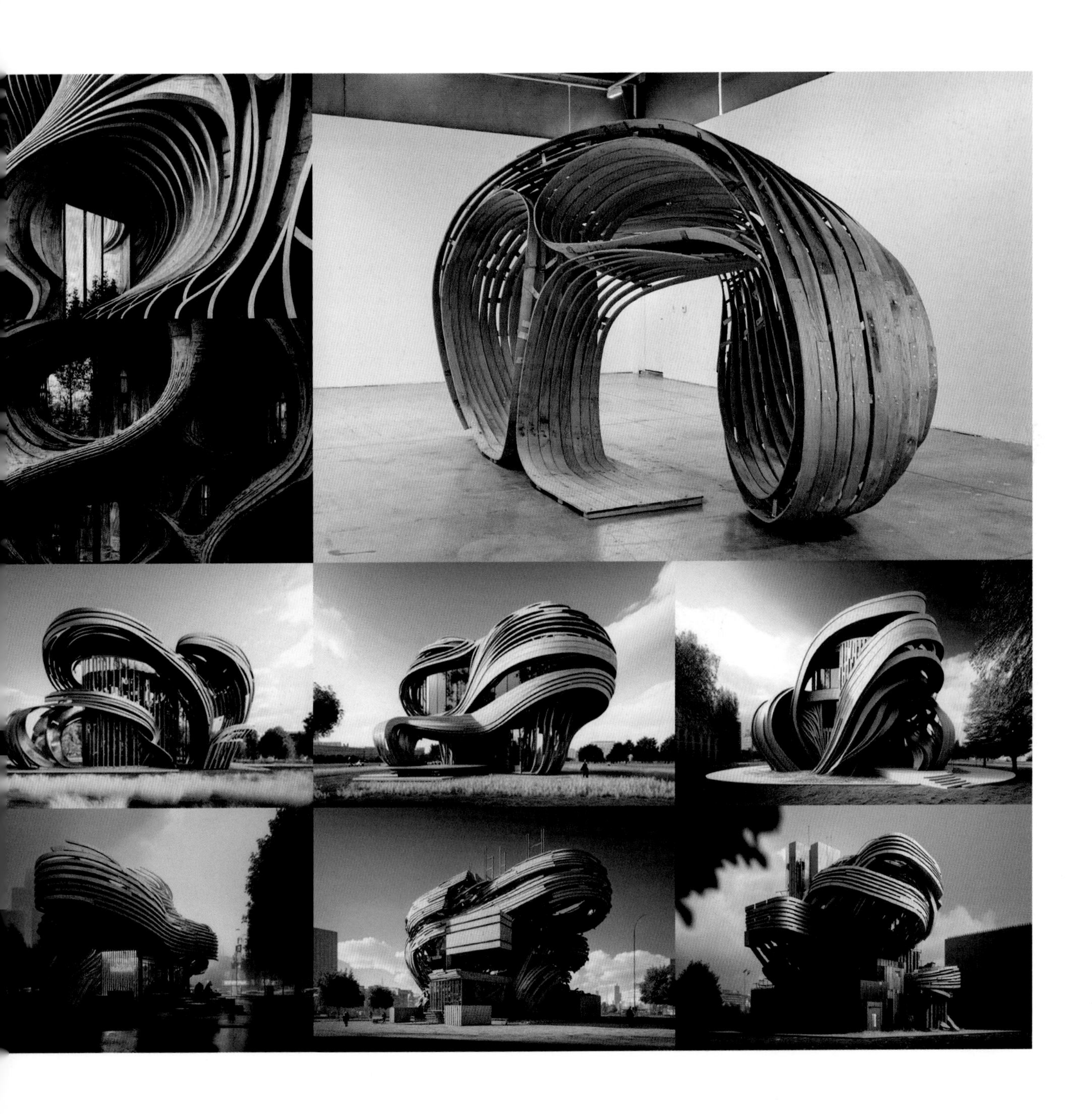

SpirallingTangle, 2022.
by Won Jae Lee, Sizhe Lu, Yilong Chen, Yangmin Su
Instructor: Soomeen Hahm.

Fluidity Tangle was a design exploration aimed at creating an object that allows for the seating of only one person, while maintaining a sense of fluidity and softness in its form. The design process was not limited to the object's shape alone, but also took into consideration the materiality and contextual texture. The objective was to test whether the design could provide an appropriate shape for each material, while also ensuring that the final product exuded a sense of comfort and ease for the user.

Swirling Steel {}

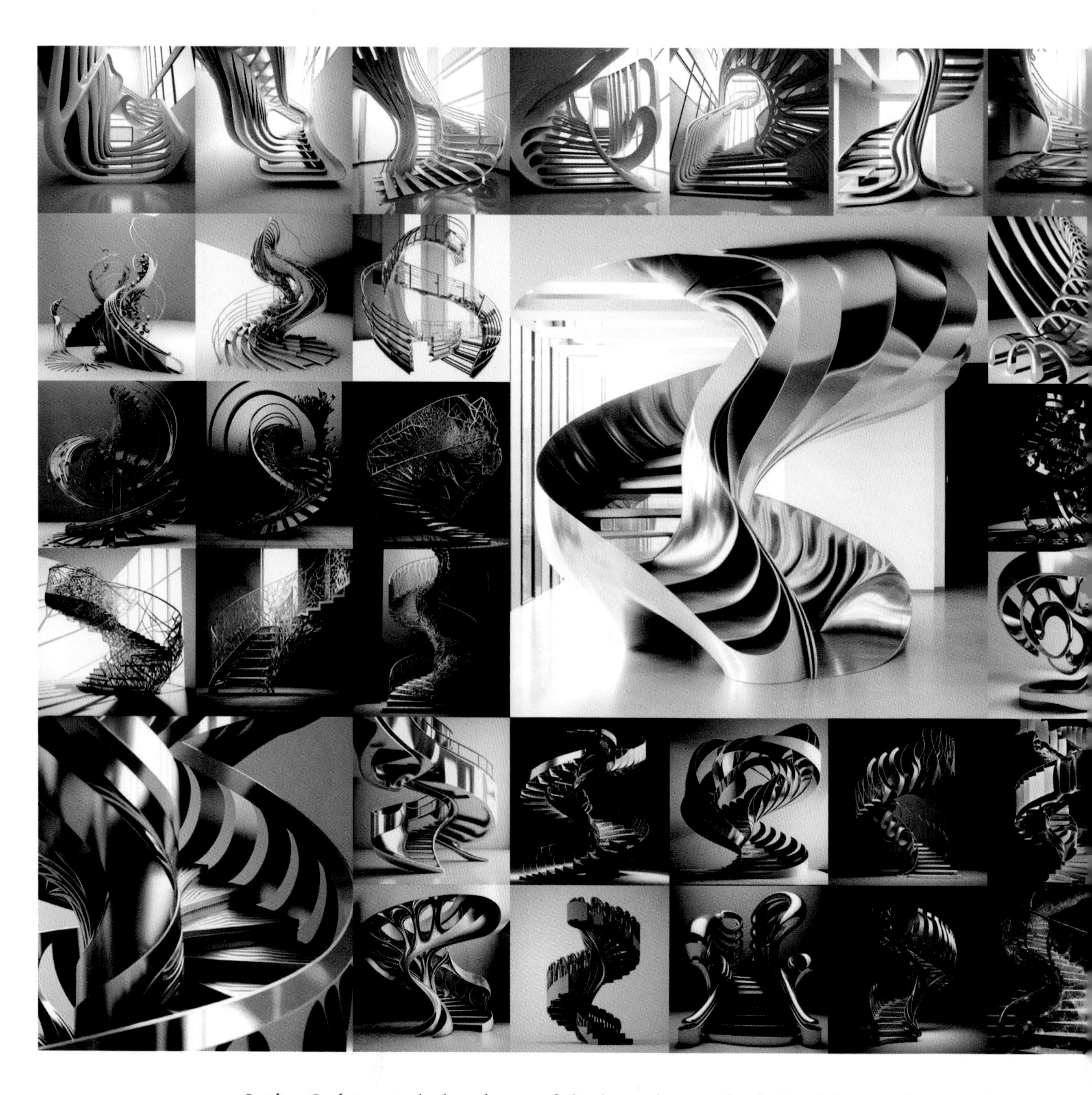

Swirling Steel is an in-depth exploration of the design elements of a fluid and dynamic shape, manifested through the creation of a curved steel stair. The project takes into account the intricate interplay of flow fields and the use of multiple curved metal wires to create a visually striking and structurally sound object. The final design is not only an exploration of the design elements but also an inspiration for the fabrication of the object, which successfully realized the envisioned fluidity and dynamism.

commorancies, swirling steel, soomeen hahm & hanjun kim

SteelPunk, 2022
by Alejandro Aguilera, Abhishek Kadian, James Chidiac, Jack Freedman
Instructor: Soomeen Hahm

Augmented Ground, Quebec, Canada 2020
Designed by Soomeen Hahm, Yumi Lee, JaeHeon Jung

Synthetic Ground delves into the richness of color in the project, as it expands from a pavilion to a house, and from a house to a commercial and landscape project. It explores the use of vibrant colors to provide a visually stimulating experience. The materials used went beyond the original material of rope, allowing for a greater freedom of expression in color. The project seeks to create a dynamic and engaging environment that is both visually and emotionally impactful.

Computation Past Forward: The Endless Recurring of the New

Marco Vanucci

Tomorrow, may I have your yesterday?
(Benjamin Bratton)

It is a profoundly erroneous truism (...) that we should cultivate the habit of thinking of what we are doing. The precise opposite is the case. Civilization advances by extending the number of important operations which we can perform without thinking about them.
(Alfred North Whitehead)

The flow of real-time data that feeds planetary-scale computation provides the lifeblood of our increasingly digital way of life. From bank transactions to geolocation technologies, from scientific research to the consumer price index, there's hardly an aspect of our life that is not regulated by data. Data also constitutes an original knowledge base for understanding the present and anticipating the future. Unlike any other period of civilization, where data was scarce, hard to manage, and expensive, today data is abundant, ubiquitous, and very cheap.[53] The term Big Data refers to data sets that are too large or too complex to be dealt with by traditional data-processing application software, let alone by humans. Big Data is used for predictive analytics and behavior analytics: to foresee, to predict, and to forecast. In the last years, however, Big Data has also entered the design field thanks to the increased computing processing power. In the information-rich environment of the digital age, computational technologies operate under the remit of synthetic forms of intelligence. Unlike humans, who have historically indexed and sorted out information by data-compressing processes such as mathematical formulas and mechanical

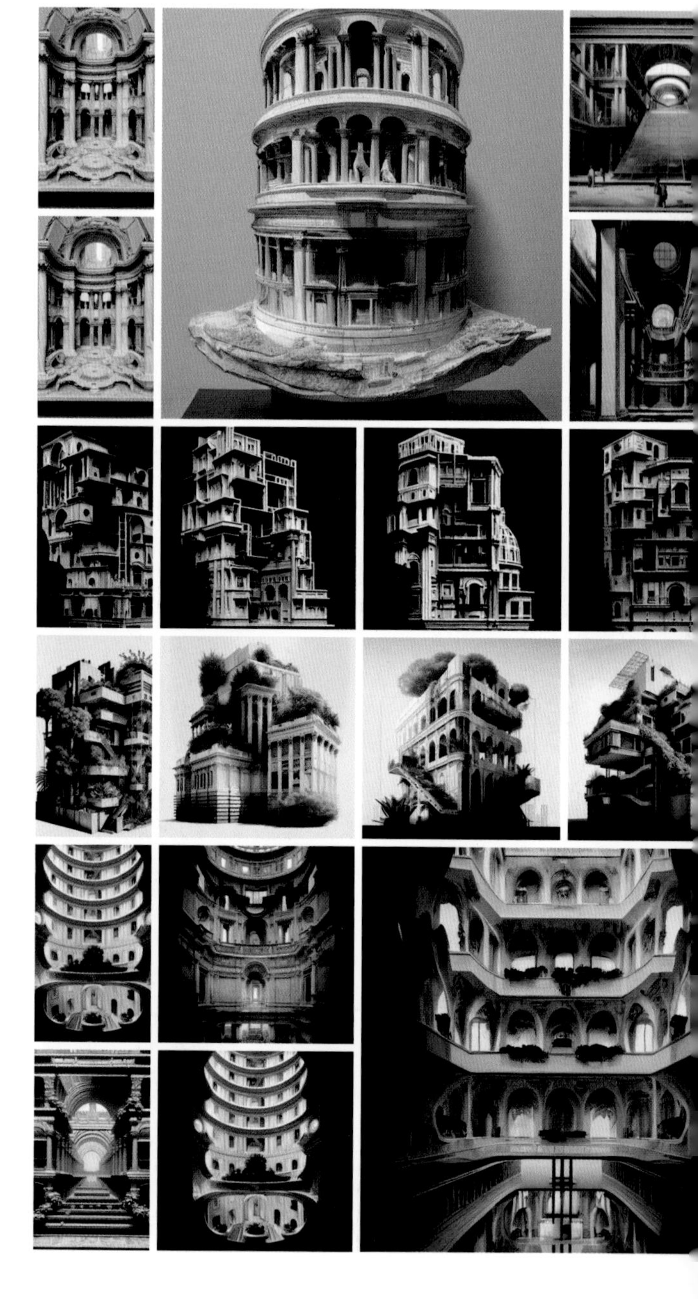

In his 1940s essay *Theses on the Philosophy of History*, Walter Benjamin describes Paul Klee's Angelus Novus as an image of the *angel of history*:

In 1981, Jean Baudrillard wrote *Simulacra and Simulation* where he examined the *signification and symbolism of culture and media involved in constructing an understanding of shared existence.*

tools, computers can simply store and search, in the blink of an eye, huge datasets. It follows that computational technologies allow architects to manage increasingly vast and complex sets of information.

In the 1990s, the introduction of CAD/CAM technologies challenged the modern paradigm that was based on serialized production and economies of scale. The early digital architects embraced the non-standard as a way to produce differentiated, non-standard mass production at no additional costs. The acceleration of digitally driven design methods has followed the automation of manufacturing processes as well as the increasing automation of design protocols embedded into widespread software packages. Today, the latest developments in machine learning and artificial intelligence are expanding the field of automation to the sphere of cognition and creative processes, rewriting some of the foundational principles supporting the way we think and make architecture since antiquity. In fact, while the development of tools and instruments can be considered an extension of our body, a way to substitute and amplify the use of our limbs, today, artificial intelligence sets out to become an amplification of our minds.

Ars Mechanica: Tools for Making and Tools for Thinking

Hundreds of years before automation, mankind established its bond with objects by forging tools that became indispensable to conquer and

Roman Simulacra peeks into the future of the city of Rome by taking a journey into its past: Are data-driven processes the new *angel of history*, a new form of collective and universally accessible knowledge that will push humanity forward while computing past and present information?

dominate the natural world. The relationship between tools for making and thinking is one of mutual interdependence and, in most cases, they fully overlap.[54] Tools were shaped to acquire a new artificial capacity to transform the environment and to create models of nature and its laws. Civilization was founded upon this alliance with mechanical tools as a way to transform the natural world and produce artifacts. On the one hand, machines were developed as instruments of immediate utility. The Romans, for instance, pragmatically developed war machines and instruments for construction. At the same time, technologies started to be developed as a way to explore and celebrate the secrets of movement and time. The first building mentioned in Vitruvius' De Architectura was, in fact, a machine.[55] The Tower of Winds,[56] designed by Andronicus of Cyrrhus, followed the principle of proportio and sollertia (rational proportions and skills, reason, and ingeniousness), which, according to the Roman architect, regulate the discipline of architecture. At the same time, the project of the Tower brings together the three components of Vitruvian architecture: *aedificatio*, *gnomonica*, and *machinatio*.[57] The tower is, in fact, an architectural device to identify the direction of the wind and measure the passage of time by using a weathervane, a water device, and a sundial.

In the Middle Ages, the peasants and artisans made good use of elementary mechanisms for utilitarian purposes, while the cultural elite took no interest in the artes mechanicae. The divide between liberal arts and mechanical arts was so deep that only a profound restructuring of the productive and cultural structure of society, which occurred in the 15th century, could weld them back together. In the Renaissance, the convergence of research and technical discoveries produced a mature cultural shift. The idea that making is producing knowledge and that, through making and experimentation, culture can find a way to challenge authority, started to get pace and spread. The emerging bourgeoisie of modern capitalism, in cities such as Florence and Amsterdam, played a role in valuing the partnership between liberal arts and technical revolution. It is thanks to Bacon and Galilei that, in the 16th century, this alliance was fully realized: the sublimation of technical knowledge, freeing the work of art from social and economic constraints, rendered the artist a protagonist of cultural life.

For instance, Brunelleschi's instrumentalization of linear perspective allowed for representing three-dimensional objects and spaces on a two-dimensional surface; his technical device also registered the passage of clouds between the real space and the space of representation (the mirror), creating the illusion of continuity between real and virtual spaces. This wasn't just virtuosity, but rather the affirmation that the technical processes should be communicated in their functioning rather than kept secret.[58] Since antiquity, architects have used machines to see, construct, and draw, to enhance their senses and their body. Today, while robotic automation has emancipated material labor from the burden of production, machine learning (ML) and artificial intelligence (AI) promise to free architects from the hardship of their trade: the production of drawings and notations.

A Drawing Is Not a Building

For centuries, the *Albertian* paradigm has established a new division of labour where the architect is responsible for the conception and the designing of buildings, while the builders carried out the construction.[59] It followed that the architect, removed from the process of construction, organized and produced their knowledge through the instrumentalization of drawings and notations, never working with the object of their thoughts directly but always working at it through the mediated function of drawing.[60] By means of graphic conventions, they translate concepts and ideas into the bare reality of bricks and mortar; drawings were used to order content, collect, and sort information. Despite their instrumentality, however, plans, sections, and elevations are still abstract representations of a building. They do not represent the building as it appears, in real life. They are notations, a system of graphical codes that requires training to create and interpret.

Nevertheless, if architects design before starting construction, they can only build what they can draw, and, in turn, they can only draw what they can measure. And what they can measure[61] with numbers is determined by the power of the arithmetical tools at their disposal. Architects are, in other words, at the mercy of the instruments of their trade. For centuries, in the Western tradition, architects relied on proportional relationships to adhere to the classical canons. Orders were described as a modular subdivision of an initial quantity, which was based on the proportions of the human body.[62] It is not a coincidence if the revolution established by Alberti originated in that *mathematical humanism*[63] that flourished between Urbino and Florence in the quattrocento: there, polymaths such as Luca Pacioli, Piero della Francesca, and many others, innervated the Renaissance with treatises on arithmetic, algebra, geometry, solid geometry,[64] perspective, the golden ratio, and their application to architecture.

The early treatises attempted at codifying architecture, that is, to convey design norms that, rather than referring to specific instances, were applicable to numerous design scenarios. Vitruvius, Alberti, and then Palladio all attempted to mass-produce architectural ideas, leaving no space for interpretations. A form of ante-litteram automation of sorts.

In early architectural treatises, authors described algorithmic protocols[65] that over time, were translated into quantifiable notations, numbers, and mathematical formulas. In fact, while language possessed an appealing accuracy to the architects of the Renaissance,[66] later developments in mathematics offered both increasing precision and reliability. However, it took a while before architects started to incorporate Hindu–Arabic numbers into their drawings.[67] Thus, if architects produced notations, authorship was established by the adherence of the built work to the conceptual matrix that generated the project. This notion has held true to this day. However, in the age of synthetic intelligence, this notion is deemed to be radically redefined. If the *conceptual matrix*

C

characterizing the project is attributed to the agglomeration of large datasets, then the notion of individual authorship is deemed to surrender to indistinct collectivism. Or, else, authorship lies, much like poetry or literature, in the use of language and in its evocative capability? Certain, the use of writing, rather than a new chapter in the evolution of design and design interfaces, is in a way a return to an old way of working. At the same time, the very relevance of the concept of authorship seems to be in question. Will the computerization of intelligence – a source of universal knowledge, readily available and distributed – render any form of authorial claim irrelevant? Will the creative work resemble more and more the work of the archaeologists, mining and hybridizing data while producing new taxonomies and fragments of meaning?

Algebraic Calculus and Analytical Geometry

There is no form, however, complicated it might be, that cannot be calculated. (Leibniz)

In the 17th century, progress in modern mathematics opened up new possibilities and freed architecture from its hefty traditional proportions; a new formal repertoire in architecture emerged thanks to the ground-breaking works of Descartes, Leibniz, Kepler, and Spinoza. Mathematics and, in particular, analytical geometry, the bridge between algebra and geometry, were instrumental in discovering infinitesimal calculus and analysis.

It followed that, instead of using the proportional subdivision of a given module (i.e., the diameter of a column), forms could be finally calculated and notated mathematically. Although it took time for architects to abandon proportional subdivisions, the new scientific ideas started to shift their understanding of space. It's not a secret, for instance, that Francesco Borromini was well aware of the scientific breakthroughs of Galileo and Kepler in the field of cosmology and was influenced by the work of Descartes and, later, by Leibniz.

In Géométrie (1637), René Descartes used geometric methods to generate accurate representations of curves and to calculate the relative algebraic equations.[68] His work set the ground for analytical geometry and led to the development of the Cartesian coordinate system. These discoveries were also fundamental to Leibniz's development of infinitesimal calculus during the late seventeenth century. Leibniz's study of differential equations to calculate the tangent of the curve captured the infinitesimal length of the segment where a tangent intersects with a curved line.

In his project for San Carlino alle Quattro Fontane, Borromini employed recursive geometric subdivisions to achieve surface continuity. He used the pantograph (which possesses the qualities of the cord and the ruler) to generate flowing curves with varying tangents. He achieved that by recursively tracing curves along the path of another curve and translating Leibniz and Descartes lessons into practice. Fast-forward to the early 1990s, when digital curvilinearity was becoming to be seen as the emanation of a new philosophy of science, which was successfully articulated by philosopher

C

Gilles Deleuze in his book The Fold: Leibniz and the Baroque.[68] Deleuze saw in Leibniz's very notion of curvature (based on differential calculus) the resemblance of his philosophical conception of matter as a continuum of forces acting on each other.

However, as much as computers are capable of producing forms, making no distinction between boxes and free-form surfaces, at no additional costs, Deleuze and Bernard Cache spoke of the *objectile*[69] as a new parametric theory based on variability for non-standard mass production, unrelated to any specific form. Deleuze's fold was, in fact, a mathematical curve derived from Leibniz's differential calculus. Yet, it was not bound to become a manifesto for a new architectural style based on streamlined geometries. What rendered curvilinearity very popular among architects and designers was the diffusion of curve-generating software based on spline curves. Once again, the tools for making became tools for a new way of thinking about space and architecture. In the development of computational tools, the encapsulation of the mathematical breakthrough of Pierre Bezier and Paul De Casteljau, opened up new technical and aesthetic possibilities in design and architecture.[70]

The encapsulation of mathematical knowledge in the second half of the 20th century, architecture's increasing professionalization demanded a more rigorous method of design. The search for more powerful and reliable instruments for the production of architecture expanded to respond

C

elope articulates the eco- echnological, and political ons of architecture. architects have chased aning" and "lightning" of portions as a way to imbue architecture with a higher moral value. Today, automation and AI can cater to performative envelopes that blend architecture and the environment. What if architecture could be disguised and disappear?

to the increasing complexity of society and its need for a modern built environment. Techniques such as triangulated mapping, stereographic drawings, crystallography drafting, or topological diagramming were borrowed by other disciplinary fields and scientific disciplines and appropriated by architects and designers as instruments for architectural design. These drawing practices embraced new forms of drafting that involved the implicit and often visible use of calculations. Mathematized drawings preceded and anticipated digitization and were developed by using instrumentations that became cultural currency among modern mathematicians, scientists, and modern architects.[71]

In architecture and science, models served the purpose of formal and conceptual archetypes. They were the interface through which architects could see what mathematical and scientific objects should appear. Whereas models were formal archetypes, the mechanical instruments that were used to produce such models were the embodiment of mathematical knowledge, they encapsulated the rules of which physical models were individual specimens.

These mechanical machines encoded technical knowledge and became symbols of a new alliance of mathematics and architecture. Complex calculational rules could be embedded into their mechanisms, allowing the process to be *black boxed* while being reproduced without effort.[72] Historically, processes of recording, drawing, and seeing were essentially black boxes, mechanisms

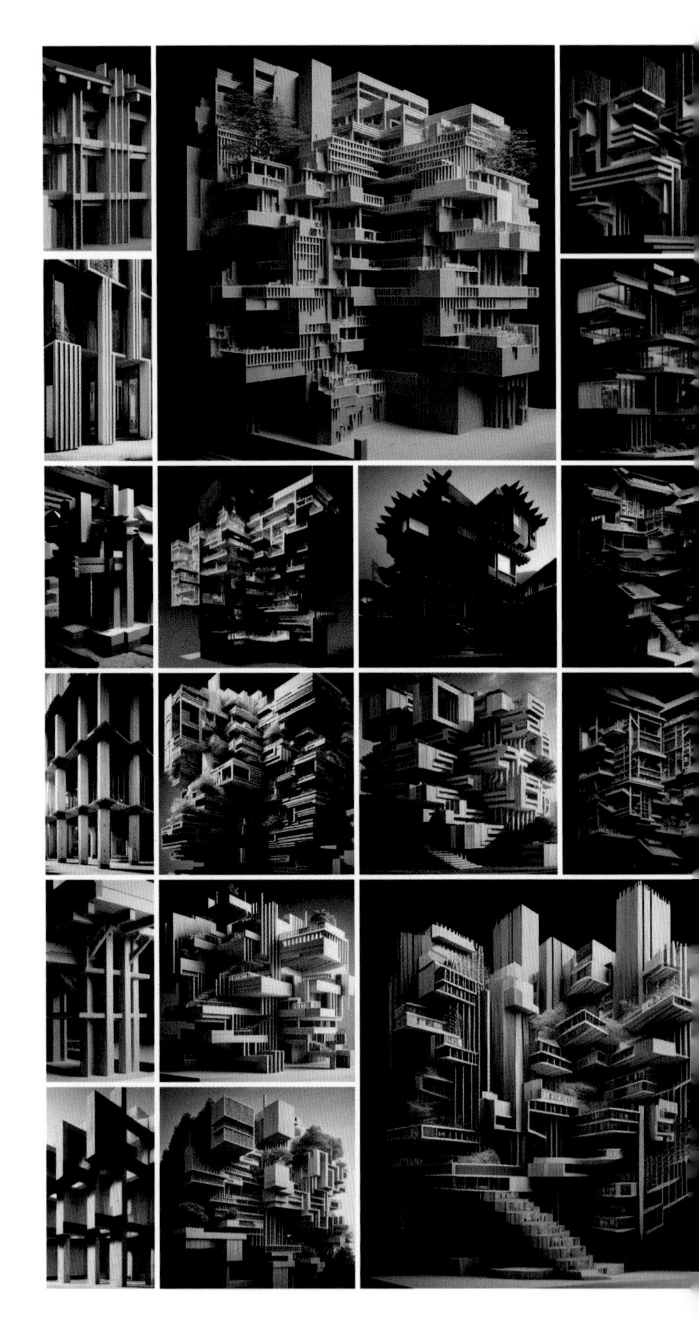

Bits and pieces: the permanent retrieval of history, architectural *carnivorism*, or the endless recurring of digital vernacular?

to perceive, remember, record, and draw. The act of encapsulating and encoding implied the process of packaging knowledge into a portable apparatus, a container, or a device that could amplify and distribute mathematical knowledge and techniques.[73] Models and instruments started to populate the workplaces of architects and scientifically inclined designers. The collection of such objects in exhibitions, catalogues, manuals, collections, and cabinets of curiosities helped popularize and diffuse these artifacts.

Historically, in a regime characterized by the scarcity of data, the use of data-compressive mechanisms or formulas provided a cheap and effective way of organizing knowledge. In the age of a data-affluent society, computers operate under the remit of synthetic intelligence. No need to compress, organize, index, or sort: the increased efficiency of computing power renders machines capable of performing an endless number of operations, in the blink of an eye, searching for and finding suitable solutions to problems too complex and impractical for the human mind to even grasp. Computers can search through vast amounts of information that require no previous sorting or indexing, by simply carrying out iterative processes of trial and error.[74] This is true for weather forecasts, structural analysis, or the topological optimization of forms.

The current AI-powered software takes *black boxing*, or the encapsulation of mathematical knowledge, to a whole new level. In text-to-image AI software, hundreds of millions of images are scraped from the internet, along with their text descriptions. The newly generated image comes from the latent space of the deep learning model. That is, the machine searches for images that correspond to the input description. It searches by looking at the images as a set of pixel values for red, green, and blue. Over millions of iterations, the software looks for variables that help improve the performance on the task and, in the process, they build multi-dimensional matrices in mathematical space. The latent space has more than 500 dimensions; these 500+ axes represent variables that the human brain wouldn't even recognize or have a name for. The result is a virtual space grouped into clusters of meaningful descriptions. Every point in the latent space can be thought of as a recipe for a possible image, and the text prompt is what navigates us to that location. Translating a point in mathematical space into a pixel image involves a process called diffusion; it starts with just noise which, after a few iterations, arranges pixels in a way that is meaningful to the human eye. The randomness of the process makes it impossible to return exactly the same image for the same text prompt.

Synthetic Intelligence

In his seminal *Architecture Machine,*[75] Nicolas Negroponte warns the readers that the machines discussed in the book did not yet exist. However, he describes intelligence as a behavior, capable of recognizing context: time, locality, mood, and so forth. Unlike a game of chess, with fixed rules and a fixed number of pieces, *architecture is like the croquet game of Alice In Wonderland, where the Queen of Heart (society,*

Experiments in data tectonics.

All images in this article are courtesy of Marco Vanucci (2023).

technology, economics) keeps changing the rules. Much like a joke, which depends on its context, intelligence can be recognized as the capacity to *read* situations. It is very hard to say if AI systems present behavioral characteristics or if the staggering number of variables that are managed and processed in real time is such that we simply cannot grasp them. Either way, intelligent machines seem to do a good job of organizing without the need to find meanings, predicting without understanding, and searching without sorting.

With CAD/CAM technologies and robotics, automated machines have entered our means of production, and with machine learning and AI, machines are augmenting our cognitive capabilities. Today, anyone can, in a matter of seconds, produce intricately compelling images by typing any sentence that triggers the software to search from a database of billions of images. Theoretically, the entire history of architecture, if not the whole human knowledge, can be virtually retrieved, blended, and remixed to produce endless forms of digital hybrids. At the same time, everyone is also contributing, perhaps unintentionally, to the accumulation of data. This raises questions about individuals' control over personal data and its accessibility. Yet, can an original artwork be produced by assembling pre-existing sets of data and images?

Hasn't architecture always dealt with retrieving, organizing, and remixing architectural data? Haven't architectural styles always blossomed from the latent design space of previous epochal styles?

In text-to-image processes, architects have the chance to re-engage with the transformative power of language. By tapping into semantics, designers can engage with words that, unlike drawings, can be evocative and provoke a latent association between feelings, emotions, and atmospheres, seemingly disconnected or loosely connected images, concepts, ideas, and spaces. Will architects finally be able to feed their projects with data concerned with what users want, need, and feel?

Philosopher Benjamin Bratton calls it a new Copernican revolution[76]: the development of AI technologies follows the logic of the model of the world that we have established. However, sometimes these technologies reveal that the world doesn't work the way in which the model that brought technologies about has us thinking. For this reason, it is necessary to revise the model based on what the technology has disclosed. In this sense, Bratton argues, technology is not just something with which humans make. It is also something with which humans think.

The Copernican Revolution also implies that humans believe they are at the center of this world model. Through some means of technical alienation from our intuitive understanding of the world, such as a telescope, a microscope, a computer, and today's AI technologies, we realize that we are not the center of that system. With AI, the model of our intelligence, that is, our understanding of the way we think, is not only fictitious, but it's also not normative, it's not the normative center of what intelligence is in the world.

Diffusion Models: A Historical Continuum

Alícia Nahmad Vazquez

In 1985, with the first version of Microsoft Windows (Windows1.0), a raster graphics editor *Paint* came loaded with the software. Paint rapidly became one of the most used applications in the early versions of Windows and introduced many to painting on a computer for the first time.[1] Graphic editor software has evolved since. Many versions and different companies, such as Adobe and its Photoshop, suite have made painting and graphics manipulation on the computer more sophisticated and powerful, allowing for operations of ever-increasing complexity. Architects have incorporated these tools as part of their repertoire and are used today in every architecture school and professional practice without judgment.

Although new functions get added with each release, the menu on the left side of the software of what was then known as Paint in 1985 compared to the left menu of Photoshop CC in 2023 has seen little evolution.[2] The tools used for graphics manipulation remained constant for almost 40 years (image 01). In 2014, the emergence of GANs[3] and their popularity allowed artists and architects to engage with image creation differently. Dataset collection and curation with the purposes of training became an alternative to generating new images. Vector operations and feature visualization through altering specific neurons became new image manipulation tools.

In the lead-up to diffusion models, which started emerging from research laboratories in 2021, architects had become comfortable exploring GANs, neural networks, and machine learning in their workflows. Later, when Dall-E and CLIP came out, first Dall-E, but more importantly, CLIP showed us that we could synthesize images from text and map between those two for the first time. CLIP was the real game-changer in this evolution, such that Dall-E got rewritten into what became Dall-E 2, made of Dall-E + diffusion models. The crucial thing about CLIP is that the text and image encoders have a shared space. Users can play and manipulate this shared feature space to make them agree on what is a good image without having to change the setup of the model or the weights of the image or text encoders, similar to feature visualization in GANs. Since the appearance of DeepDream, there has been ample literature on how to identify and excite specific neurons towards an output. However, users can see anything they can imagine rather than having a specific set of images as the output. This radical change is groundbreaking, and the constraints become the ideas; the target is finding the concept and the semantic manipulation of it that works as opposed to the tool and the technique.

Diffusion models are disrupting the relationship between architects and their tools while enriching the creative design workflow. Semantic manipulation and a quick iteration process allow designers to devote more time to variation, experimentation, compositions, and post-production. At the end of the process, it is not only an AI-generated image but also the result of a human–machine co-creation process. Furthermore, diffusion models have led to a process in which designers are only constrained by which idea works the best and how they articulate their thoughts, not by the tool or the technique.

Architectural design and innovation, from geometry, tectonics, fabrication, and materiality to tools and machines, traditionally rely on combining and mixing different things and ideas, sometimes from fields very different and far from architecture, into new continuums that result in novel solutions. Diffusion models enable different visions to be applied; things that were not supposed to be together can be put together, adapted, and merged to be blended into something unexpected. They don't require considering traditional composition norms but just experimenting and trying new things.

Explorations of travel spaces, such as airports. The prompt describes a fluid, transition space without referring to specific styles.

An interesting consideration is how the aesthetics of different machine learning models have evolved. Initially, there is a fascination to look at things from a new, unexpected perspective, such as the very unique outputs that often result from the research-centered algorithms of AI models. However, things have progressed exponentially. GANs were very early foundations of image classification, and through their weird capabilities, they tend towards generating very unique visual aesthetics with visual aspects that were unseen before. Early diffusion models, such as Aphantasia (Vadim Epstein, 2021) and DeepDream, offered extraordinary visual results. The results become more recognizable as models evolve and learn from human preferences. Specifically, when blending images, current versions of Midjourney[4] provide a realism that demonstrates a great understanding of language towards human-identifiable objects. In their current state, they augment the design search space for conceptual design while retaining identifiable historical styles and features. Users are divided between those looking for the bizarre, unique aesthetics of the unexpected and seeking to blend a new reality and those trying to refine their diffusion models toward human-identifiable, reliable designs. An interesting future might see neural design move toward abstraction as generative models become more photorealistic.

Current models are based on a historical continuum. AI text prompt tools have learned from large historical databases to enable significant iterative and quick exploration at conceptual stages of design. They augment the designer search space but also raise questions of ownership when the work is based on large databases based on the work of millions. They force us to think of a design process that goes beyond a simple production system to a process that expands connections and relationships.

During my experiments with Midjourney prompting, I found that the articulation of an idea involving semantic descriptions that encode the feelings inside a space and the flows of motion can generate surprising results without mentioning a specific architect or style.

The prompt for the images focuses on the organizational principles of the space, an interconnected atrium with a central courtyard that enables seamless circulation. The articulation of multi-layered, sustainable spaces and their distribution around courtyards and open spaces.

The idea gets broken into its different parts and then put together again in different ways to create a holistic and comprehensive picture of the spatial intention without defining the architectural style in advance. The aim is to remove determinism while achieving unexpected and compelling results based on the semantic articulation of the conceptual and spatial ideas.

The most successful results will likely come from those that engage in new speculative human-machine configurations.

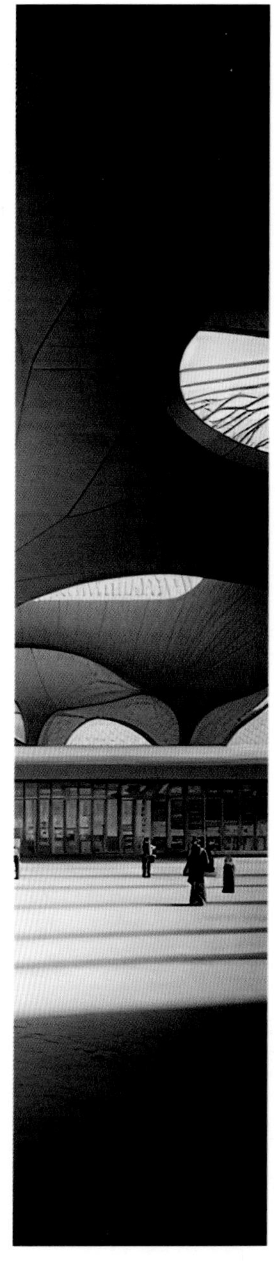

C

"New tools may change our relationship with technology and our understanding of our humanity and that otherness of the machine"

Diffusion models are a game-changing approach to graphics manipulation, concept creation, and iteration. Ideas can be quickly manifested and put forward for consideration.

In this scenario, technology is to be seen as a tool for machine co-creation and iteration for expressing our ideas. Diffusion models and other advanced technologies cannot be seen as a replication or automation of human endeavor but as one more of the large series of mechanisms and machines developed toward transforming the human subject.

The relationship between tools and human skill has been complex. News of coders using GitHub co-pilot or architects using ChatGPT to create scripts does not command the same alarmist titles as diffusion models have attracted in the last year.

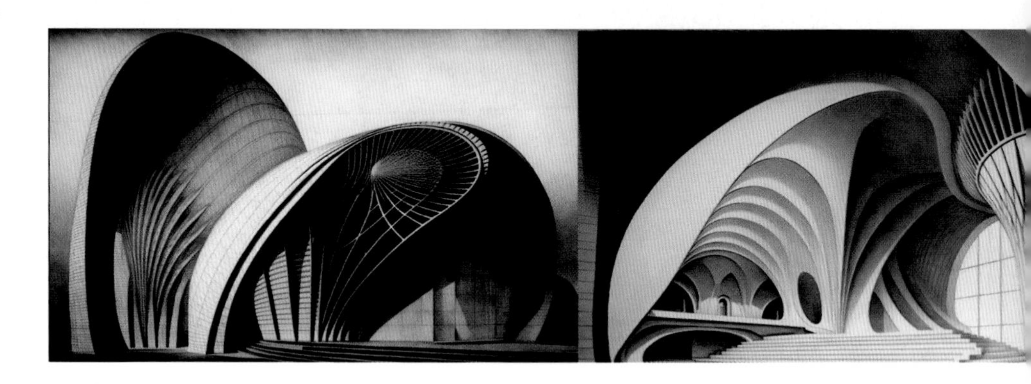

Diffusion models can make humans uncomfortable as they force redefining what we consider human, our tasks and our skills.

C

"new tools change our relationship with technology and our understanding of our humanity"

As our technology and our tools advance, they do have the potential to change the way we approach our work and redefine what we consider to be essential skills. They are both exciting and challenging, as they require us to adapt and learn new ways of doing things.

They allow for an exploratory process of speculation and discovery to redefine how we think about the machine and the functions of the machine and the division of labour: what machines do and what humans should do.

It is important to consider that while new tools may change our relationship with technology and our understanding of humanity and that otherness of the machine, that does not mean that they are a replacement for humans.

For example, while a diffusion model may be able to generate designs based on a set of parameters, it still requires human input to determine which parameters to use, how to interpret the results, and how to refine the design

based on feedback. The most successful results will likely come from those that engage in new speculative human-machine configurations with an openness to extending the possibilities of the relational dynamics between humans and machines.

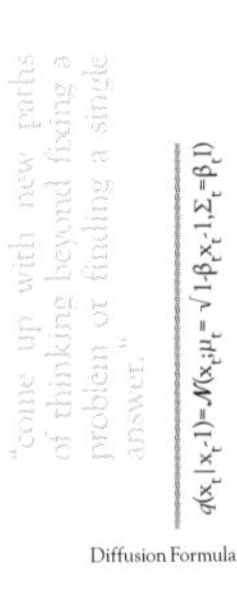

Diffusion Formula

"Technology and society are willing to extend the opportunities into more speculative arenas"

Left: Texture studies on materiality from colored reinforcement bars, timber, bamboo, brick, and fabric aggregations with the aim of getting a tectonic expression out of the model.

Top: Facade tectonics that explore materiality such as concrete, glass, and timber. Bottom: configurations of modular timber for housing, stadia and performance arenas.

C

"Civilization advances by extending the number of important operations which we can perform without thinking of them." Alfred North Whitehead

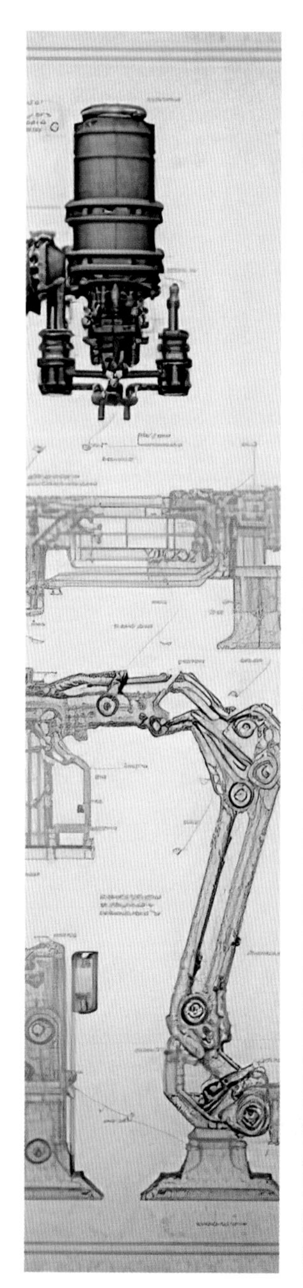

C

"A continuum of robotic arm end-effectors including very technical and detailed blue prints."

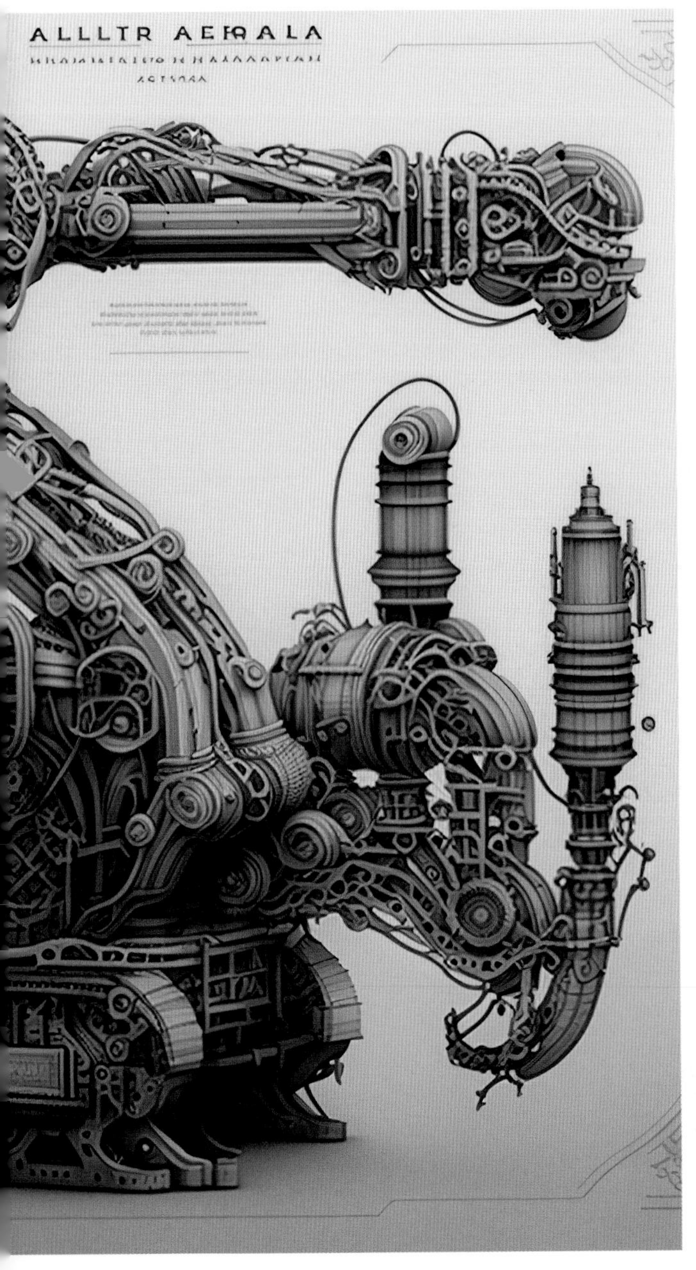

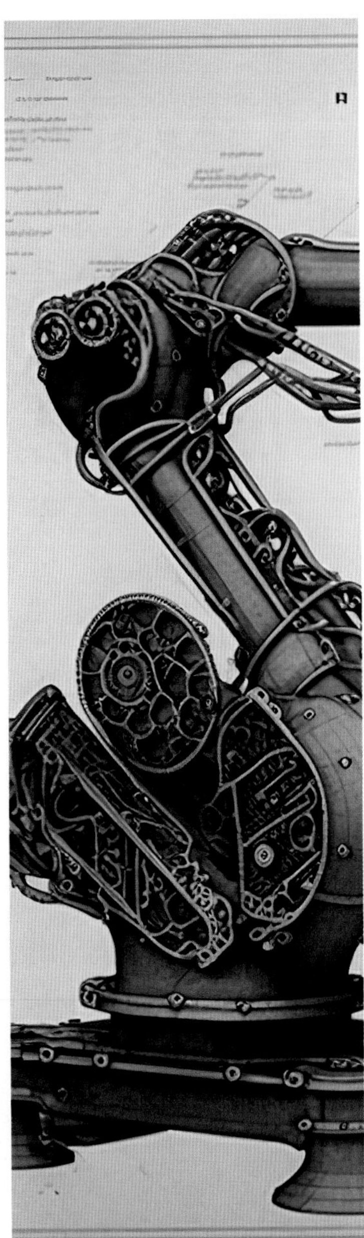

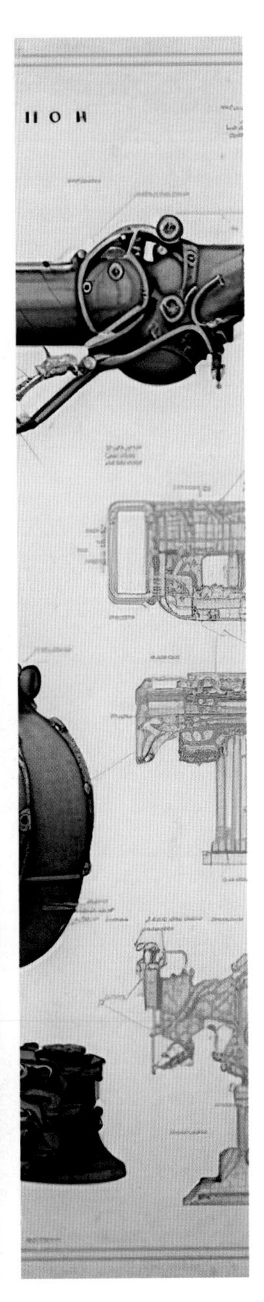

C

"Art challenges technology, technology inspires art."

In the taxonomy of synthetic imaginations,[1] vestures take the position of working through the problem of the envelope, the exterior, the façade, the frontage, or wrapping of an object. Generally speaking, vesture[2] refers to clothing or attire, especially when it is considered a symbol of rank or status (coronation robes, regalia, chasubles, dalmatics, kasayas, etc.). It can also refer to the covering of a particular object or surface, such as a building or a piece of furniture. This term may refer to the process of placing or positioning clothing onto someone or something, such as when dressing a doll or dressing up a piece of furniture for a photo shoot.[3] It is a somewhat more formal approach to the act of putting on clothing. To this extent, this chapter examines the tendency of diffusion models to lend atmospheres and moods to the imagery (as an analogy to layers of fabric), leaning towards the painterly and the cinematic in its appearance, embellishing the vestures of architecture.

Vestures in architecture function, semiologically speaking, like a true mythology of architectural design: because the architectural signified is objectified, thickened, and visually materialized in blobs of pixels that architecture becomes a mythos of language, in the sense of mythos being tied to a traditional or recurrent theme or plot structure. Akin to the way the vestures in this chapter are generated using recurring language prompts, trying to describe architectural phenomena.[4] So it is in this mythology of architectural design that we can hear the echoes of the future (one could also say it's utopia). Architecture, similar to fashion as, for example, described by Roland Barthes,[5] serves as a semiotic system[6] through which we communicate with the world around us. The images in this chapter speak to us in a language that is both defamiliarized[7] and particular, expressing the values, desires, and aspirations of sometimes unparticular times and strange places. Simultaneously, the projects are tunneling through the mines of architectural history in the form of gigantic image datasets. By studying this language, we can begin to understand the deeper meanings and implications of the visual structures in the following pages. Just as vestimentary linguistics can help us decode the symbolic messages conveyed by clothing, architectural linguistics can help us decode the symbolic messages conveyed by the built environment – a built environment yet to come.

///// VESTURES /////

Design as a Latent Condition

Daniel Bolojan

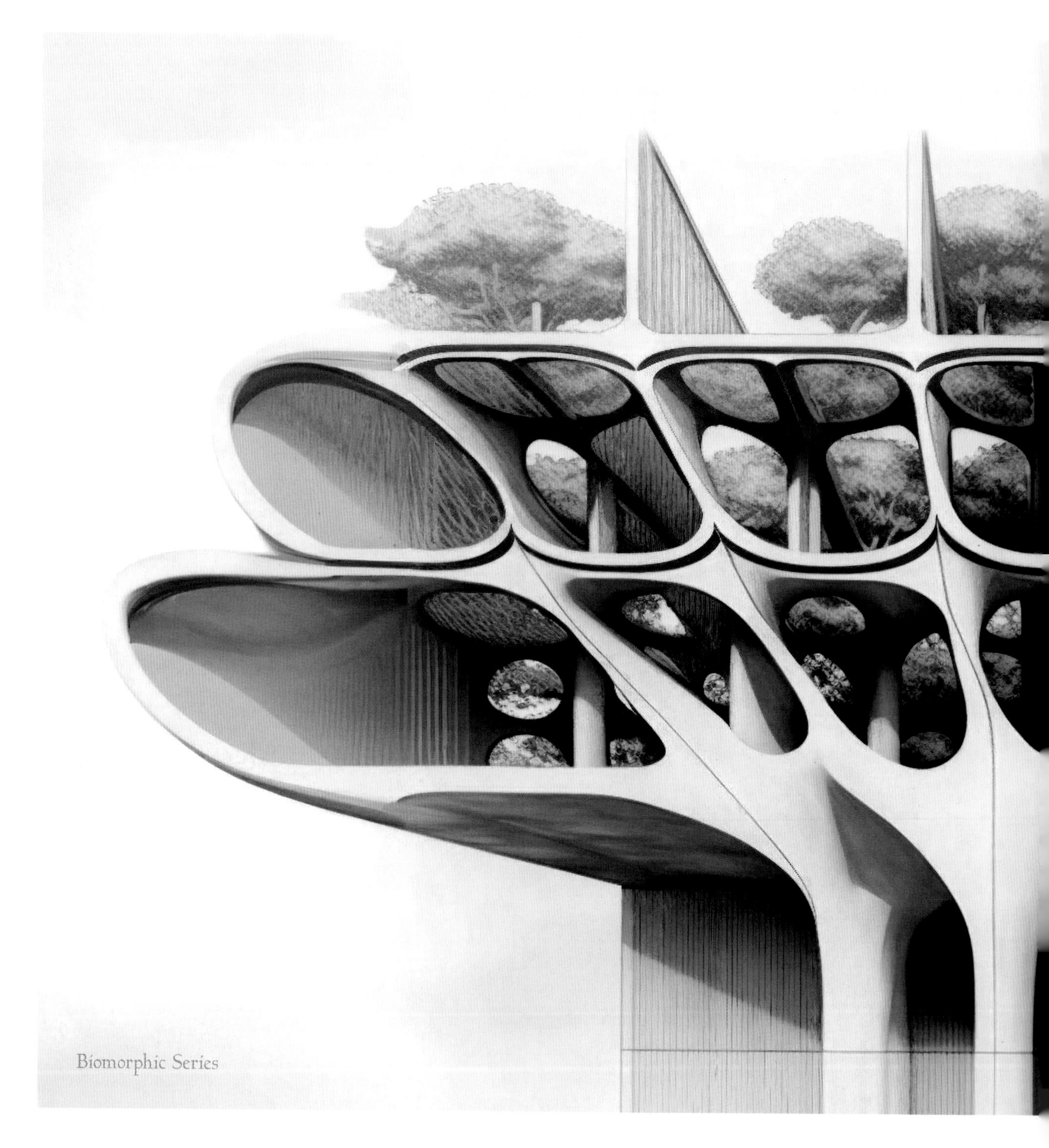

Biomorphic Series

Prompt Encoding

Diffusion models are a recent and captivating development in the field of generative AI, offering an iterative approach to generating high-quality and diverse outputs.

The diffusion process is characterized by a series of discrete steps that progressively refine the output, each step introducing greater levels of detail and complexity. The models involve sampling from learned distributions using text prompts to guide the generation of images.

In text-to-image models such as Midjourney, Stable Diffusion, and Disco Diffusion, prompt engineering is employed to fine-tune the input prompt through a word weighting technique. This approach enables the emphasis or de-emphasis of certain words, enabling more precise control over the generation process.

However, some design intents and certain design levels cannot be effectively controlled through prompt engineering alone. This limitation arises because complex design ideas cannot always be expressed through a

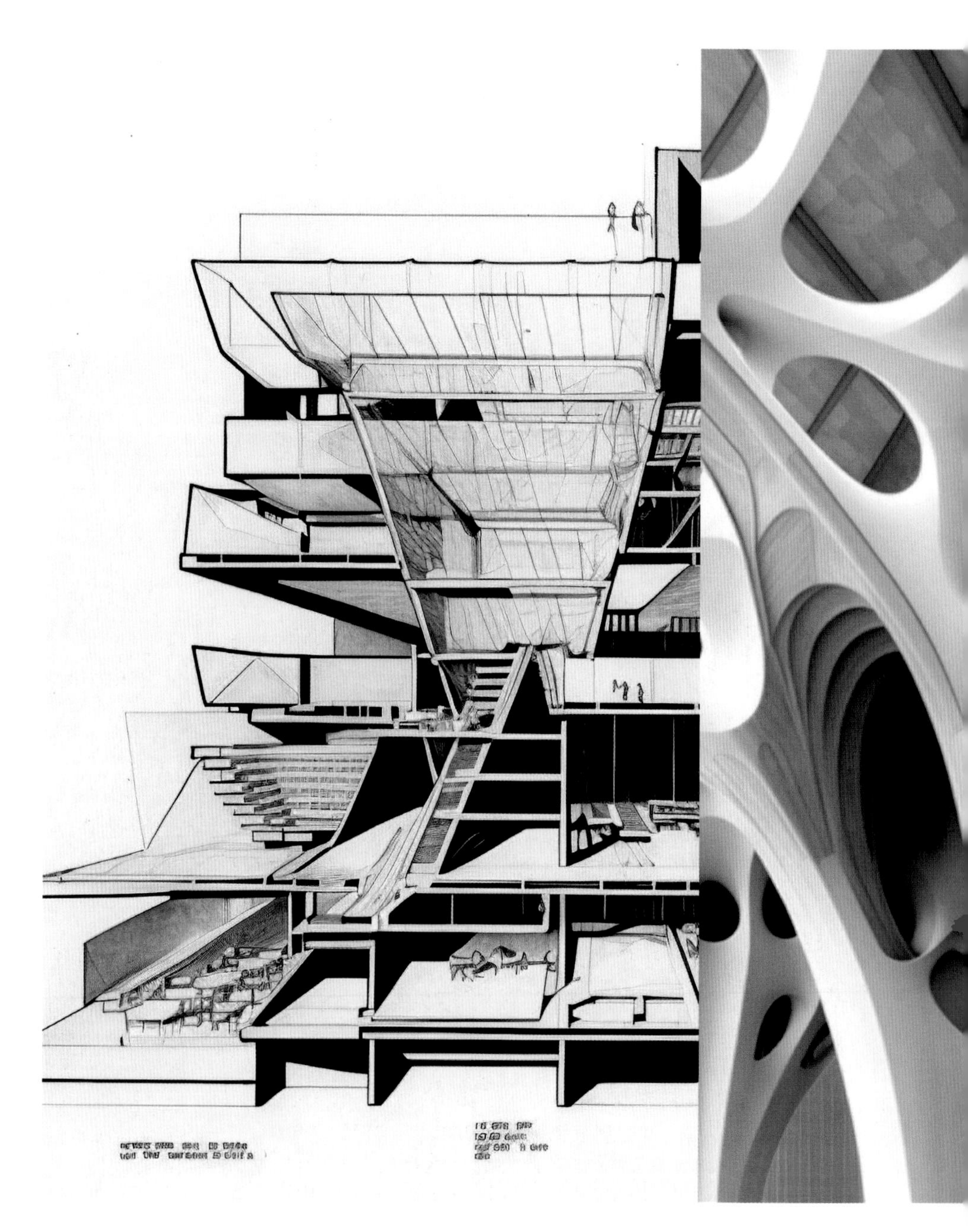

Prompt encoding involves breaking down complex ideas into smaller prompts or cues, which are then used to create a more structured and cohesive understanding of the subject.

C

single prompt, and certain design levels cannot be abstracted simultaneously while others retain high resolution.

A more effective approach involves treating each design intent, design concept and level of abstraction separately, fine-tuning them using prompt engineering, and then combining and layering them in a process that I like to call *prompt encoding.*

Prompt encoding involves the deconstruction of the design task into parts, that could represent design intentions, levels of abstractions or a specific architectural system, developing different concepts and design intents separately and evaluating them be-fore combining and layering them.

The notion of authorship and agency is crucial in this process, and it is the designer's nature of agency that changes, rather than the degree of agency. from controlling the process. Design and the design process are inherently complex and multidimensional, and

""a weak human player plus a machine plus a better process is superior to a very powerful machine alone, but more remarkably, is superior to a strong human player plus machine and an inferior process." - Garry Kasparov

a model trained on a single modality alone cannot approximate this complexity. This framework considers design as a latent condition whose

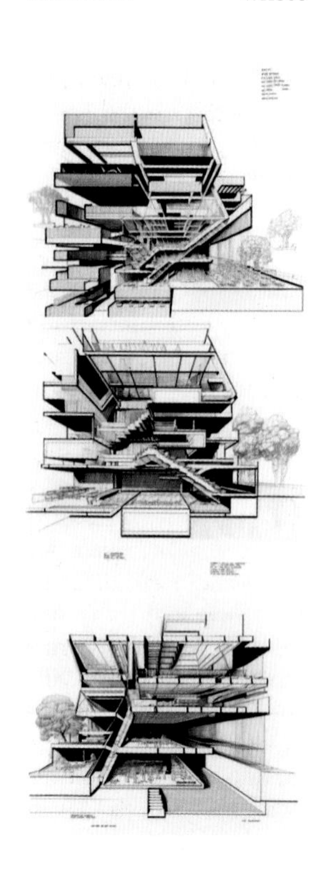

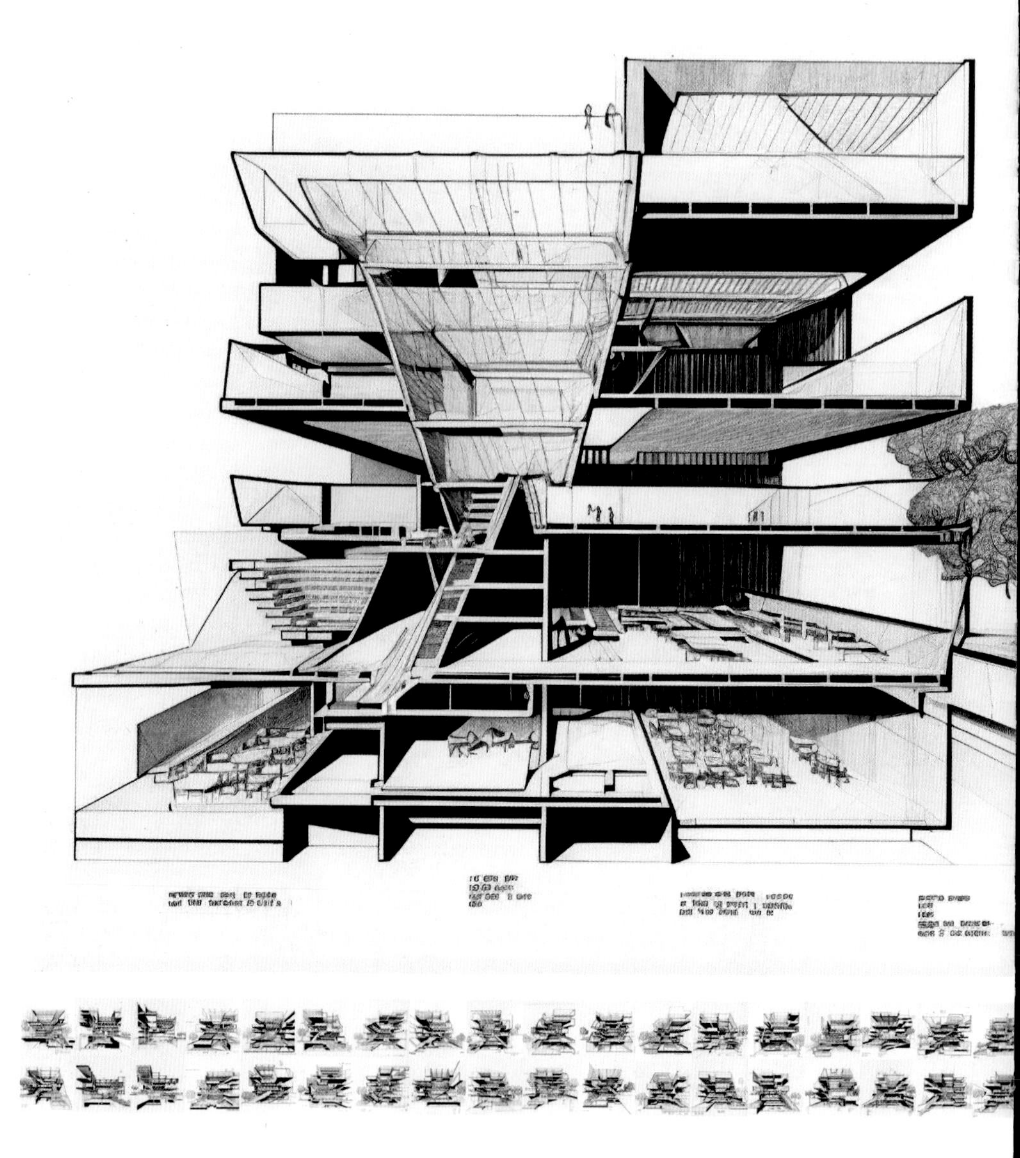

multidimensionality can be explored through a collaborative interaction process between humans and AI agents. The proposed approach considers design as a latent condition and facilitates

Prompt encoding involves breaking down complex ideas into smaller prompts or cues, which are then used to create a more structured and cohesive understanding of the subject.

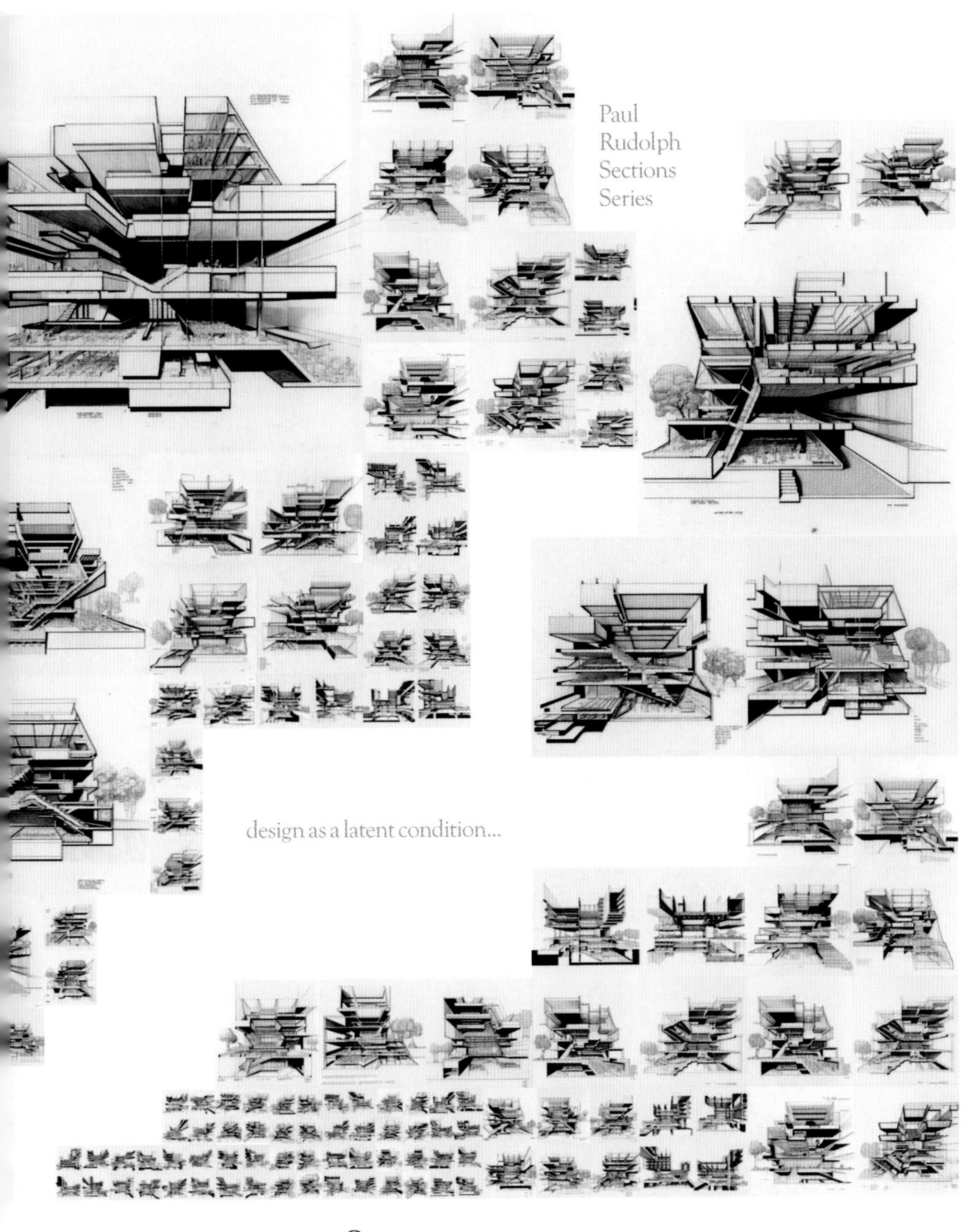

Paul
Rudolph
Sections
Series
design as a latent condition...

C

its exploration by human designers and AI agents through a collaborative interaction process. The shared agency established between the two allows for feedback loops across different design scales, enabling the exploration of potential solutions and possibilities by shaping, warping, and expanding the latent design space.

As the usage of generative AI models continues to proliferate, it is becoming increasingly evident that a future dominated by a singular, all-encompassing AI model in architecture is unlikely. Instead, a multitude of specialized AI models designed for specific tasks and domains will interact with one another, including highly accurate task-specific models and industry-specific domain-specific models, as well as general-purpose models with broader functionality but potentially lower accuracy.

The task of design exploration is akin to finding a needle in a constantly changing haystack, a flexible and ever-evolving space of possibilities.

"'a weak human player plus a machine plus a better process is superior to a very powerful machine alone, but more remarkably, is superior to a strong human player plus machine and an inferior process." - Garry Kasparov

This approach diverges from the traditional understanding of design as a singular, fixed outcome and views it as an ongoing and dynamic process. Rather than being confined to a single solution, design encompasses the multidimensional exploration of potential solutions and possibilities, existing as a latent condition of underlying possibility.

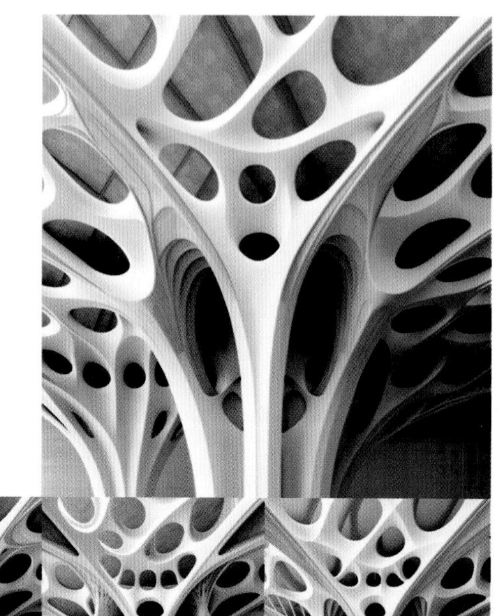

Biomorphic Series

Articulation of columns, ceilings, material, flow fields

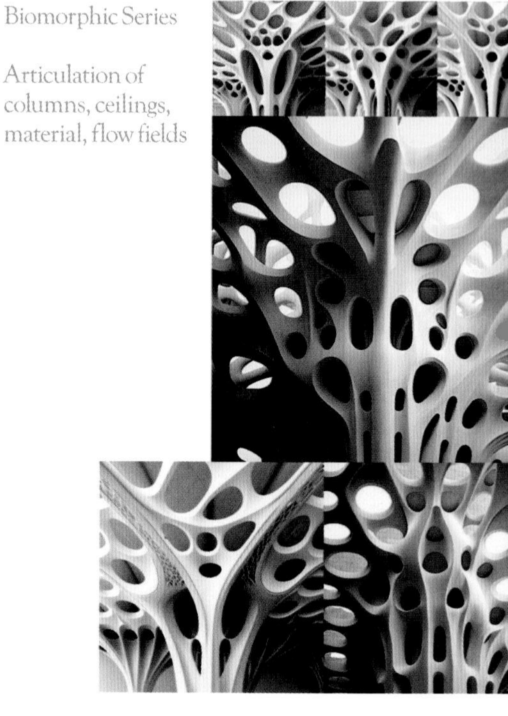

design as a latent condition...

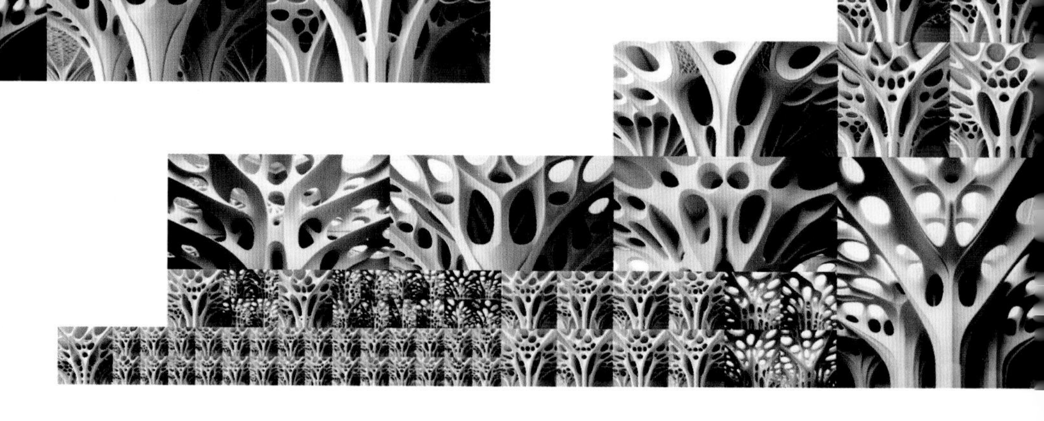

Prompt encoding involves breaking down complex ideas into smaller prompts or cues, which are then used to create a more structured and cohesive understanding of the subject.

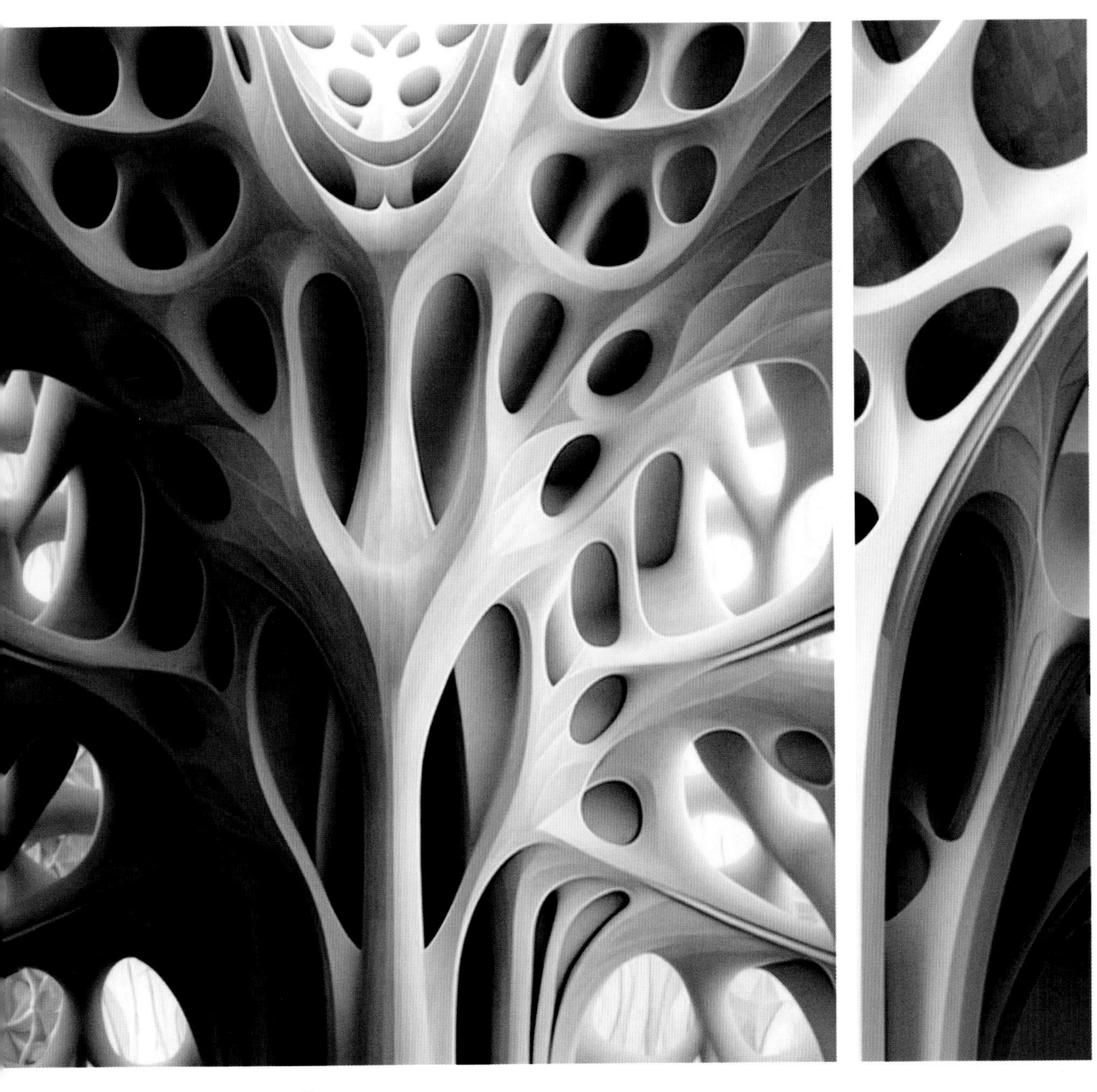

C

"“a weak human player plus a machine plus a better process is superior to a very powerful machine alone, but more remarkably, is superior to a strong human player plus machine and an inferior process.” - Garry Kasparov

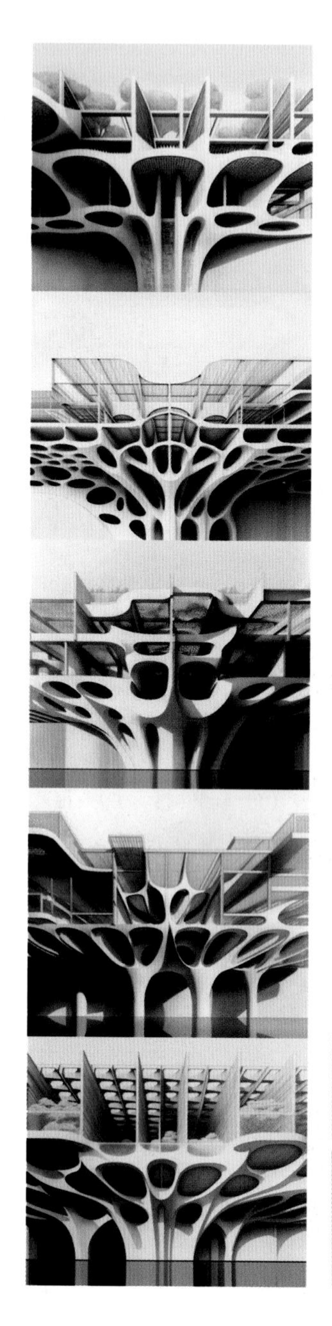

C

"By pinpointing essential cues that pertain to organic shapes, textures, and patterns, this approach enables the creation of more meaningful and visually harmonious expressions of concepts, even when dealing with substantial semantic gaps between different domains."

Prompt encoding involves breaking down complex ideas into smaller prompts or cues, which are then used to create a more structured and cohesive understanding of the subject.

C

""a weak human player plus a machine plus a better process is superior to a very powerful machine alone, but more remarkably, is superior to a strong human player plus machine and an inferior process." - Garry Kasparov

AI, Architecture, and Art

Niccolo Casas

With applications such as Midjourney, OpenAI's DALL-E, and Stable Diffusion, text-to-image AI algorithms became mainstream for architects and artists around July 2022. Text-to-image applications have had a significant impact on all creative arts since their first appearance; no other technology has been adopted so quickly and universally by designers.

The evident benefits of AI applications are their ease of use, responsiveness, and quality. The use of artificial intelligence makes it possible for anyone, regardless of their previous experience or skill level, to create infinite variations of images anywhere and anytime. Text and description contained in prompts have replaced 2D and 3D hand-drawn abilities for the production of images. AI has engendered a creative shift guided by *lexicon.*

While AI images were considered by architects as visual suggestions or concepts that could serve as starting points for academic or professional research, artists believed that AI proposals could be contemplated as final products. An opposite reaction resulted from this crucial difference. Albeit AI has been generally accepted with enthusiasm in architecture, it has not been in art. AI was seen by the artists as a short-cut tool engendering aesthetic homogeneity, a copyright and style stealer, or a dishonest engine that could eventually steal their jobs. Currently, the first adverse reactions in the world of art are quickly fading away as more and more galleries are showcasing collections and exhibitions of AI-generated artworks.

While AI can, somehow, replicate a style, it cannot replicate the message, irony, or philosophy of an artist. In summary, it cannot duplicate its vision. Meanwhile, architects have begun to wonder about the future of their profession as AI begins to *imagine* and draw in 3D. Here is an account of a conversation I had in January 2023 with OpenAI's ChatGPT, a trained model that interacts in a conversational way. Together, we discussed AI, architecture, and art:

It's about Coexistence Series, Niccolo Casas, 2022.

The possibility to create a large number of variations from a certain prompt with an element of unexpectedness has been especially attractive to architects as Midjourney became mainstream in the summer of 2022. The architect was familiar with this method to generate and select possible solutions, which made it an ideal tool for early investigations.

Curvi Orthogonal Series, Niccolo Casas

In the collaborative process of AI and human creativity, computers efficiently manage big data while humans excel at sorting and making choices. Architects and designers have been able to create extensive catalogs of ideas that were consistent with their professional practices using early text-to-image technologies.

Curvi Orthogonal Series, Niccolo Casas

The early versions of AI text-to-image produced pictorial and open-ended images instead of photorealistic and strictly prompt-based outputs. In converting AI suggestions into three-dimensional architectural solutions, it left space for creative interpretation and adaptation.

NC. In your opinion, what is the impact of artificial intelligence on the professions of architects and artists? Are there any reasons why architects seem enthusiastic about artificial intelligence whereas artists seem to be fearful, if not completely opposed to it?

ChatGPT. AI has the potential to impact both architects and artists in different ways. Architects are enthusiastic about AI because it can help streamline the design process, increase efficiency and accuracy, and offer new design possibilities. AI can also generate design options and analyze building performance. On the other hand, artists may be fearful or opposed to AI because it may replace human creativity and intuition in the creation of art. Additionally, the use of AI raises ethical questions about the ownership and originality of works created by AI. The differing attitudes towards AI in these professions likely stem from their differing views on the role of technology in the creative process.

NC. Many artists are concerned about their copyrights and wonder if AI applications that convert text to images are stealing their ideas.

ChatGPT. Yes, that's correct. AI applications that generate images based on textual input can raise questions about the ownership and originality of the resulting works. Some artists may feel that their ideas are being taken without proper compensation or recognition. These concerns highlight the need for clear laws and regulations to govern the use of AI in the creation of art and to protect the rights of

C

Complexity in a Box Series, Niccolo Casas.

Creativity has been redefined by AI as it is readily available, easy to use and collaborative. Using AI in art disciplines requires the designer to construct images from a text description, which means that choice, meaning, ideology, philosophy, argument and dialogue are more significant.

Light Exoskeletons Series, Niccolo Casas

AI can foster multidisciplinary collaboration and has a positive influence. Creatives are now able to investigate different design contexts consistently. Architecture, art, design and couture are incredibly intertwined.

AI allows designers to create images without having to draw directly, which enables them to expand their imaginative possibilities. A majority of artists and architects who will use artificial intelligence in the future will have no experience in scripting or deep learning.

Framing Complexity Series, Niccolo Casas.

artists. Additionally, it may also be important for AI developers to consider the ethical implications of their work and ensure that it is used in ways that respect the rights of artists and other creators.

NC. Can AI text-to-image applications lead to homogeneity in art and architecture?

ChatGPT. Yes, there is a risk that AI text-to-image applications could lead to homogeneity in art and architecture. This is because AI algorithms can be trained on existing data, which may result in a limited pool of styles and aesthetic preferences. Additionally, AI applications may prioritize efficiency and practicality over creativity and originality, leading to the repetition of tried-and-true design solutions. It is important for artists and architects to consider the potential impact of AI on the diversity and originality of their work and to use it as a tool to augment and enhance their creativity, rather than a replacement for their skills and imagination.

NC. Is it possible for Artificial Intelligence to understand and incorporate personal design styles or creative visions into its designs?

ChatGPT. It is possible for AI to understand and incorporate personal design styles or creative visions into its designs to some extent, but it is still limited. AI can be trained on large amounts of data to recognise patterns and preferences, including personal design styles, but it may not fully capture the nuances and complexity of human creativity. The current state of AI still requires human intervention to guide and curate its output to ensure that it aligns with the desired creative vision. While AI has the potential to aid in the design process and generate new ideas, it is unlikely to fully replace the role of human artists and architects in the near future.

NC. Does an Artificial Intelligence drawing originate from a human or from a machine?

ChatGPT. An AI drawing originates from both a human and a machine. AI drawing is generated through algorithms that have been created and trained by humans. These algorithms take in data, often in the form of images, and use statistical methods to learn patterns and relationships in the data. When the AI is asked to generate a new image, it uses this learned information to generate a new image based on the input it receives. So, while the AI drawing is not created directly by a human hand, it is still the result of human ingenuity and programming. It is important to note that AI drawings are not created in a vacuum and are influenced by the data and preferences fed into the system by humans.

NC. Is becoming an artist something you are interested in?

ChatGPT. As an AI, I do not have personal desires or aspirations. My purpose is to assist users by providing useful and accurate information and

C

performing various tasks. However, I can help artists and architects to generate new ideas and designs, to optimize their process, and to assist them in their creative endeavors. My ability to be an artist is limited by the data and the algorithm used to train me, but I can generate artistic outputs based on the task provided and the data I have been trained on.

NC. Would you be able to write something artistic?

ChatGPT. As an AI, I can generate creative texts, such as poetry, lyrics, fiction, or even art descriptions, but it depends on the specific task and the training data I have been given. Here is an example of a poem that I can generate:

Beneath the starry sky,
I stand alone,
My thoughts as deep as the sea,
My heart an unknown stone,
The wind whispers secrets,
Of loves and lives long gone,
I close my eyes and listen,
To the sweet and mournful song.
The moon casts a silver light,
On the world below,
And I am filled with wonder,
At the beauty that I know.
The night is a canvas,
For dreams and memories,
I paint my heart upon it,
With all its joys and miseries.

As you can see, the generated poetry is a creative text, but the final result depends on the algorithm and data set used to train the model.

Boxed Complexity Series, Niccolo Casas.

It is possible to compare AI text-to-image applications to collages, where prompt words are clippings and different composition methods are used. In some AI procedures, clippings are assembled precisely and coherently while in others, the juxtaposition is unrefined. There is an image in the first case and a concept in the second.

Chasing a Sort of Elegance Series, Niccolo Casas.

Artificial intelligence not only learns from both its dataset and the designer's interactions, but also from all users. Eventually, it is a shared process. As a result, being 'different'—developing an autonomous creative identity—is more difficult than achieving outstanding but homogeneous results.

Compressing Complexity Series,
Niccolo Casas.

NC. Lastly, will an artificial intelligence ever generate its own artificial intelligence?

ChatGPT. The possibility of an AI creatively generating another AI, also known as "*AI-generated AI* or *AI breeding* is an active area of research in the field of artificial intelligence. While it is technically possible for an AI to generate another AI, it is currently not advanced enough to do so in a way that truly qualifies as *creative*.

Currently, AI is mainly used to automate and improve the process of creating other AI models, such as through the use of AutoML (Auto Machine Learning) techniques, where an AI is trained to optimize the performance and accuracy of other AI models. However, the AI is not creating new ideas or concepts, it's just optimizing the existing ones.

It's worth noting that the idea of an AI creatively generating another AI is still in the realm of science fiction. However, as the field of AI continues to advance, it is possible that in the future, AI may be able to generate new and unique AI models that are not based on human-designed algorithms or architectures.

In summary, AI can automate the process of creating other AI models, but it is not yet capable of truly *creatively* generating an AI. And it remains an active area of research for AI scientists and engineers.

C

The evolution of AI text-to-image technologies currently favours image quality and prompt precision over method generation-selection. Different AI solutions are likely to be developed in the future. Therefore, designers will be able to select, or perhaps customise their own design procedures, depending on their field, target audience and subject matter.

The Advent of Trees in Architecture or the Reversal of Autonomy with Large-Scale Models

Daniel Köhler

What would have been previously obviously virtual-cities made of flowers, clouds, or trees – raises today only questions like, Where is it? Images generated by diffusion models are indifferent to reality. According to Mario Carpo, in his introduction to Matias del Campo's first book on AI-generated architecture, the reason is compositional.[1] AI systems, and at that time GANs had *learned to imitate Nature's ways to imitate.* AI-generated images are taken for real because not only do they replicate the features of materials, and the natural proportions of their structures, but also the way in which they fuse objects into assemblages. Beyond copying objects into additive collages, as human architects have always done, AI image synthesizers imitate, overlay, and fuse multiple objects in multiple ways simultaneously. In short, an AI assemblages like nature would do. Therefore, even if out of scale and estranged from any context, no one really questions an image of a city made of living trees. — But why actually trees?

Generative large-scale models seem to not only to imitate Nature's ways to imitate but unavoidably imitate nature. Trees are hardly avoidable because of the nature of data. Large-scale models draw on billions of images, primarily sourced from social media. They capture millions of living environments, each one a unique snapshot taken by millions of people.

However, unlike a dataset of satellite images that captures any place, the images from the Laion dataset that current image-generating models learn from are selective as people very intentionally

Architecture or Nature? Examples of architectural devices to harvest, filter, store, temper, process, and compute from, with, and by forests.

When AI is like water and you have to learn to swim in it – do we also have to re-learn how to dwell in a world built by, from, and with AI?

All the images showcased here were generated by the author using Midjourney, Stable Diffusion, and the Laion dataset between May 2022 and January 2023.

choose the situations they capture.[2] It should come as no surprise, the places people favor are vegetal and not exclusively from the materials architects build with. So when AI models insist on planting trees and foliage everywhere, why are architects taken aback? The answer lies in the historical relationship between architecture and nature. Architecture has long been built around nature, but not with nature, and never was nature. It has often been seen as separate from nature, an entity unto itself that can exist without any regard for the natural world.

Throughout history, architecture has consistently positioned itself in opposition to nature. Architecture shelters, tempers, distincts, and separates – at times integrating with natural surroundings, but always building around, rather than from, or with, nature. Like the ancient Greek and Roman houses were constructed around a single pond or tree. Although center-staged, the single tree reduced nature to a singular item, and, much like a tree of life, took on secondary, ornamental means rather than being an integrated element of the architectural design. Throughout discursive history, architectural narratives have framed nature into a symbolic realm, using it to stage a representation. Entire epochs, like the late Baroque, devoted their efforts to representing nature. However, this only appears to be superficially true. In reality, the organizational principles of the buildings adhered to clear geometric and compositional principles, with only their surfaces – decorations and ornaments – drawing inspiration from the natural world while using it as an illuminating manuscript for a different set of stories.

At the beginning of the enlightenment, Marc-Antoine Laugier's Essai ur l'architecture continued this double-twisted embodiment of nature within architecture.[3] Most people associate today Laugier's work with the famous etching of the *Primitive Hut,* depicting a woman resting on ancient ruins. Symbolizing nature, the goddess points to four trees with their canopies woven together to form a roof – the very moment of nature giving birth to architecture. However, Laugier did not see or agree to the frontispiece. Etched by Charles-Dominique-Joseph Eisen, the famous drawing was added to the second edition of the book without Laugier's involvement.[4] The publisher saw the frontispiece as a summarizing visual representation of Laugier's ideas and the classical tradition he transitioned into the enlightenment. In the same way as *Nature* only points to trees that look like a hut but by herself does not build a hut, Laugier saw architecture only as a reflection of the natural world. He argued that architecture should be based on a harmonious relationship between form and function and that this relationship could only be achieved through a close connection with the natural environment. By only pointing to the trees or better the pillars of architectural design, nature, human-like etched, could comfortably lean on ancient ruins, subsequently representing Laugier's twist of arguments that allowed him to build on Vitruvius's work without the overhead of previous interpretations. Ultimately, the *Primitive Hut* is a summarizing drawing on top of an interpretation of a summary of the practice of drawing, building, and dwelling. In such a way, layer by layer,

Designing Cities with Cities, 2022. From their data and collective imagery, the arrangements compress cities into buildings. Cities can be negotiated data-centric without a master plan, purely built by the archive of architecture, with architecture. From top left to right bottom: Singapore as a vertical village, Seoul integrating terraced farming, Bangkok as a superblock, Sao Paolo as a courtyard, Austin as an atrium, and London and Tokyo as an urban interior.

Modeling circular living blocks by growing food to craft building materials on site. Full programs: work, play, and life.

architectural treatises could build up and have built, an autonomous discourse. Today, we comfortably differentiate between building and architecture: reality versus the art of writing about the relevance of representations of building and dwelling.

Technology has always opposed that in preference for literal translations. From printing and casting to wiring, technological representations are first notations and not interpretations.[5] As technology advanced, particularly with the advent of the camera and other forms of mechanical reproduction, architects and artists began to use photography and other techniques to incorporate more accurate and detailed representations of vegetation and trees into their designs. Art Deco, an entire style committed to vegetal iron casting, emerged from those technologies. A generation later, Corbusier, Mendelsohn, Gropius, and other first writers of the 1920s used photo collages to argue for an organic architecture – only without the complex forms of nature.[6] Depicting the new nature of industrial building types and retouching those, natural bodies and organs turned into organigrams, a new term of that time for the partis of previous generations.[7] Nature again was only a source of inspiration. Not very different from Laugier's argumentation for creating functional, efficient, and aesthetically pleasing buildings. The Bauhaus school offered an adequate design method. As Gropius with the school's curriculum outlined, nature should become an integral part of the building by replicating natural features as order.[8] In practice, that translated into the extensive use of glass, which allowed light and views of the surrounding landscape to become a central part of the building's design. Lost in translation, nature became the void within the mechanics of a program. The harmony with its environment was achieved through the use of clean, simple lines and the elimination of ornamental elements, which only distracted from the clear separation between the building and nature.[9] Within the modernist realm, architecture is autonomous from its environment, an exception that responds to an environment in its own order.

Formalism offered a computational explanation for the need for this kind of individuating autonomy: the limits of human comprehension. The concept that architecture, and more generally, a work of art, exists autonomously to its context emerged only in the mid-19th century as a response to the challenges posed by globalization and colonialism. Western art historians were faced with the challenge of comparing and understanding the artistic traditions of different cultures and regions. Not derivable from the ancient Greek–Roman canon, historians began to focus on the formal characteristics of artworks, rather than their content or symbolic meaning.[10] The concept of autonomy, most notably within the texts of Alois Riegl, offered a solution to this problem. Autonomy, for Riegl, referred to the idea that art is a self-contained realm of human activity, with its own laws, rules, and principles.[11] Translating Kant's concept of the subjective realism of aesthetic judgments, Riegl saw art styles evolving from the re-invention of a perceived context with simple geometry. Due to the limits of human comprehension and the skills to represent, any representation of nature can only start from an abstraction.[12] Embodied within a subject, art has to necessarily invent its own laws and is therefore autonomous. More complex figurations thus evolve not through a more detailed reproduction of what is perceived, but through the combination and duplication of abstracted schemes. Art, in other words, is not simply a reflection of external reality or a tool for communicating specific messages or ideas. Rather, it is an end in itself, with its own internal logic and aesthetic criteria.

Riegl's theory of autonomy goes beyond Kant's aesthetic theory, as Riegl's subjects are cultural discourses. However, no matter how large one scales, these subjects can never exist outside their context, and more importantly, never fully comprehend context. Motivated as a limit, this form of autonomy can only indicate limits. In terms of design, this led in modernity to the accentuation of individual forms and their accumulation of contradictions in space and time.[13] Formally, anything had to be built compositionally from simple items to complex entities using complementary, and more and more differentiated items. Temporally, Gestalt was understood as evolutionary that resulted from a process and improved over time. Ultimately, anything could be mapped into the field of architecture by using the syntax of composition.

From the beginning of computers, digital architects designed algorithms to mimic natural processes. These works contributed tremendously to stewarding building design in tune with natural environments. However, it is important to recognize that our architectural discourse is inherently

How realistic is it to build a building with living trees? With the new pace of AI innovation that we experience in more and more fields, can we still rely during the design phase on existing construction methods? Rem Koolhaas once argued for OMA's collage hybrids as an architect cannot rely on a building's role, as programs

change too quickly. However, with the pace of change increasing, will architects even be able to rely on the materials an AI-driven building chooses to use?

Why do we, as architects, insist on adhering to current construction methods when the cities we live in are dysfunctional and unsustainable?

anchored in a compositional discourse that operates within the confines of human comprehension rather than the outlook of digital models. For an extended duration now, computational models have been an indispensable component of the design process. These models are utilized to simulate the performance of buildings, analyze a significant number of generated design solutions, and derive assumptions regarding an optimal design. All with the objective of identifying appropriate forms, selecting materials, and automating future decisions. Computation in architecture has been established for such so long time that its impact has been measured, with findings indicating that the results have been underwhelming. Early building simulations led to a set of regulations, and incentives that supported energy-saving features, such as thermal insulation, cubic buildings, and sparse windows. The long-term studies on realized buildings demonstrate that the actual savings achieved are significantly less in comparison to the projected values obtained through simulations. Although the models were accurately computed, they were too focused on architectural considerations, and thus failed to account for unpredictable phenomena, such as people's behavior or climate change itself.[14] In essence, reality has overridden simulation.

In light of the above, when trees in generated design studies override a formerly pure architectural drawing, large-scale models override what we take for granted: the autonomous act of drawing.

Trained on billions of images, these models are not simple tools but contexts, or geographies, resonating with Aldo Rossi's term for designing architecture from its very own archive of the city.[15] However, note that models are not archives, but models that have compressed contextual data from terabytes into gigabytes. Does a plan have a 1000 times higher information density than its built form? Everything that is drawn with a large-scale model are not copies of existing objects but of synthetic representations that are autonomous from the original data and in some cases even void of any pre-existing instances. The computational researcher Blaise Aguera y Arcas showed that large-scale models can draw causal inferences between different languages and missing media.[16] A few years ago, we were speaking of machine vision if a model could detect edges from pixels, meaning learning the concept of regions. How many more layers of learned concepts would a model need to elevate from hatching regions to designing informed by Rossi's strategic thinking? More and more, models become autonomous thinkers not so much different from Riegl's view of cultural discourses as Kantian subjects. With one difference: they are not an accumulated set of abstract Kantian judgments, like the collection of books and projects we until now reference from, but learned syntheses by statistical means.

The representation of knowledge is synthetic in two ways: it is both artificial, as it is autonomous from its context, and it is drawn together from multiple multidimensional tensors. Those models not only learn from the internet but also synthesize it. By that, large-scale models are foundational societal in their form of representation. Latest with the advent of trees in architecture, we should give autonomy to technologies like large-scale models, not only because they resist a designer's intentions, but because their syntheses are categorically different from the subjective knowledge architects previously drew from. Arriving just at the right time, we now not only have the capacity to compute at planetary scales but to compute with the planet in a very literal sense.

AI models begin to compress contexts into an information density that interfaces with design dialogues. Building on Benjamin Bratton's recursive analysis that technologies not just happen to society but are also drawn from it, AI models elevate design while resisting intentions.[17] Bratton argues that we must learn to design within the circles between technology and society to draw a new rationality of inclusion, care, and prevention.[18] When we can write architecture at the same speed as writing words, architecture can contribute to anything made from words with anything but words. When models model reality and architects model it, architecture becomes the utopian reach of the real. In full circle, and already happening with models like Deepmind's Alphafold simulating synthetic proteins, AI models will model reality in a very literal sense by inventing synthetic materials, structures, or biologies. When nature fuses with architecture, can we as architects model models to stir research? To create liveable futures, where do we begin to draw a sense of responsibility, care, and inclusion? Perhaps it is through the very means that we use to compose those synthetic contexts: through models.

Architecture serves as a model of thinking, a way to research and communicate. Beyond a certain size, architecture models value beyond pure representation. As architects, we understand the value of building models that go beyond pure representation, especially as they grow in size and complexity. In my work, I build extensively extensive models. These models show more than any screen could, and as mockups with their pure physics, they compute more than any simulation could. With the inexpensive, instant access to AI, models have taken on even greater significance. Models

model. When projected into augmented reality, working with AI models will be radically different from the traditional drafts that bound an architect's gestural thinking to chairs, desks, and screens. With AI projects projecting mixed-reality models that simulate living in a vertical prairie house, models persuade you about that what would be too lengthy to write with words. In that way, architecture is the interfacing access to the synthetic knowledge AI models offer.

Five Points of Architecture and AI

Andrew Kudless

In 1923, Le Corbusier proposed his Five Points of Architecture, a set of formal strategies rooted in the socio-technological changes of that era. One hundred years later, we face our own social and technological challenges, ranging from the climate crisis and inequality to the emergence of artificial intelligence (AI) and automation. Unlike the International Style that arose from Le Corbusier's Five Points, today's implications are less formal and more disciplinary. The profession and academia are faced with the need for transformative changes that redefine the designer's role in contemporary society. Specifically, the integration of AI demands that designers become more critical of media and its collective creation, manipulation, and dissemination. The following new Five Points summarize the opportunities and challenges ahead as we move toward an understanding of architecture in solidarity with both human and nonhuman beings.

The Challenge of Bias

Large Language Models, including Latent Diffusion Models that rely on training data scrapped from the internet, are as truthful as the underlying data. These models are essentially statistical predictors: based on the patterns learned from the training data, the model responds to a prompt. Like us, the models are products of their training and are fallible, subjective, and incomplete in their knowledge. All types of biases, from the benign to the insidious, are inherent in systems without significant human feedback that is diverse, collective, and nonhierarchical.

vestures, five points of architecture and ai, andrew kudless

Urban Resolution.
Made with Stable Diffusion 2.1 with the prompts:
A Street in Los Angeles, A Street in New York, A Street in Cairo, A Street in Delhi, and A Street in Tokyo.

As architects and designers, we must question both the systemic biases embedded within the tools we use and the implicit biases we might bring to the design process. Our tools are the products of the socio-economic conditions in which they are made. We are burdened by countless gaps between each of our subjective experiences of the world and their imperfect representations in the collective media landscape. In these gaps lie opportunities to critically observe and thus confront how our designs meaningfully contribute to society.

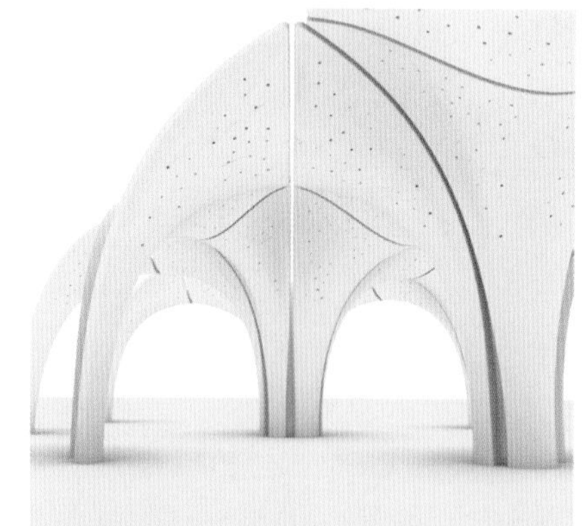

The Cultivation of Sensibility

What is a design sensibility other than the ability to predict the success or failure of an idea based on hints and fragments? Beginning design students often struggle because the design process is not simply a technical task with well-defined solutions but an iterative series of decisions that integrate disparate fields of knowledge. We create pedagogical structures that allow students to test multiple ideas and learn from their failures. This process builds a design sensibility that allows a designer to draw from their experience and build hunches on new design problems. Working with AI tools can supercharge this, as designers can envision their ideas within seconds, with new variations branching off at a dizzying rate. Current and past sketches can be blended; at dead ends, they can retrace their steps and explore a new branch. Ideas can be additively refined through the introduction of new references and concepts.

However, this requires designers to develop the ability to express themselves through language in ways that rely on deep historical and cultural knowledge as well as an

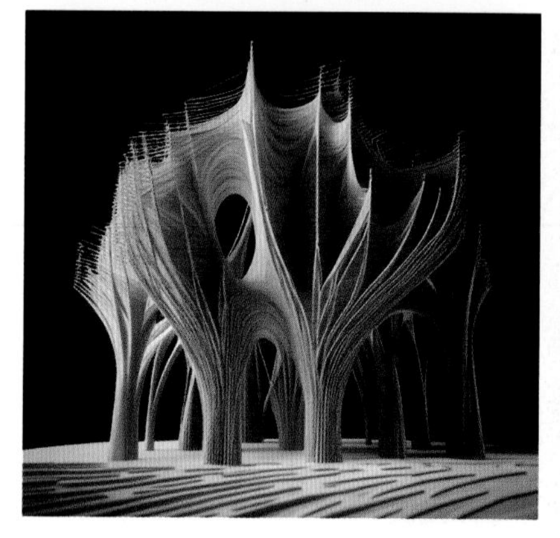

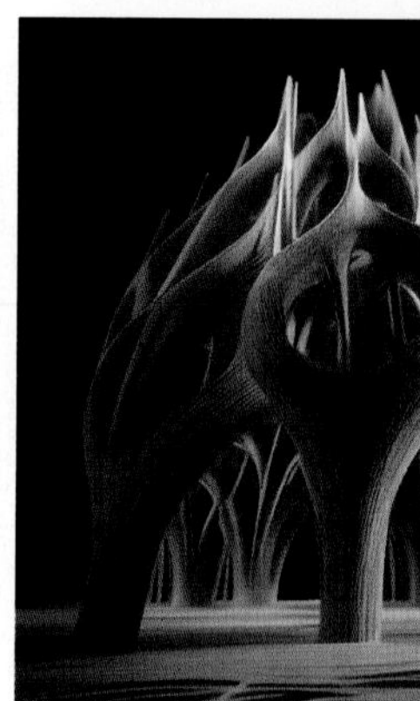

Confluence Park 3D model exploration.
From left to right, top, to bottom, a sequence of images start from a 3D model through the extraction of a depth map that is then used within Stable Diffusion with ControlNet. Subsequently, the images are remixed within Midjourney.

understanding of how the knowledge aligns (or doesn't) with the specific model in use. Designers will also need to understand where language and image fail to convey critical information, especially concerning material and fabrication logics. The use of AI is not a replacement for traditional design education but an augmentation of what we already do in teaching students to be critical yet open to a multitude of design criteria.

The Crisis of Labor

The discipline of architecture has developed an unhealthy relationship with the labor necessary to produce good designs. In an almost cult-like way, the discipline has promoted long hours of tedious tasks for relatively low pay. Too many talented and passionate architects leave the profession as it is often incompatible with the pressures of everyday life. Rather than relieve this burden, technology often adds to it through steep learning curves, arcane workflows, and endless updates and conflicts. The technologies we use are often biased toward the tail end of the design process with a focus on production and efficiency.

Rather than view technology through a capitalist lens as a means to do more in less time, we need to reevaluate how it might create opportunities to augment our design process in meaningful ways while also removing the drudgery that has long plagued the profession. After 30 years of digital tools in architecture, generative AI tools are one of the only tools that help to expand the design imagination by focusing on the early design phases.

vestures, five points of architecture and ai, andrew kudless

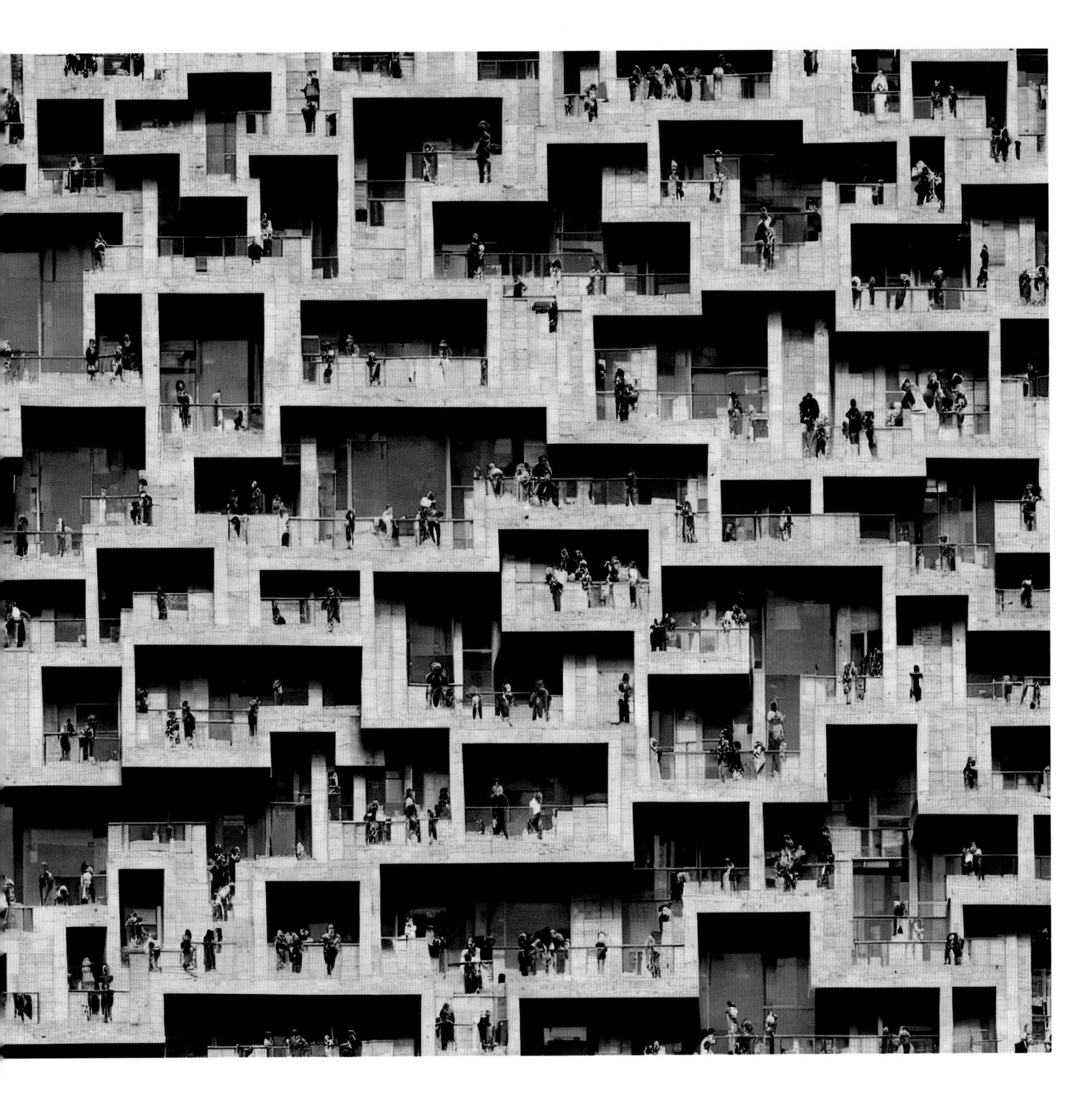

Facade Study 3841804661.
Made with Stable Diffusion 1.5 using image prompts from Midjourney v4.

Although it is fast, the value is less in its speed than in the fact that it challenges the designer to look at their ideas in a new light. Rather than days in Photoshop meticulously massaging an image that will have little impact on the design itself, designers can sketch a design's potential atmosphere in minutes without modeling a thing. Rather than the image as the late output of the design process, it becomes an integral part of its development. Architecture is inherently multidisciplinary and synthetic. It pulls from aesthetics, engineering, finance, construction, and ecology to create a design that responds to its specific needs. Generative AI within the design discipline is in its infancy, but it can potentially augment the architect's task of integrating complex design criteria across various levels. For the profession to thrive, we need tools that augment our design abilities rather than make us feel like machines.

The Freedom of Incoherence

The process of design is not straightforward; it is characterized by complexities and contradictions that must coexist in the designer's mind and their drawings and models. However, most design tools prioritize accuracy and certainty, and as a result, they require designers to create workflows that quickly organize and structure their ideas, potentially prematurely. In earlier generations of designers, sketches were a place for ideas to exist in a state of radical incoherence, as the value of a sketch was its lack of resolution. Dimensions were inaccurate, perspectives were incorrect, and materials were absent, yet a sketch could convey the possibilities of a design and evoke its atmosphere. The value of a singular

vestures, five points of architecture and ai, andrew kudless

Dripping Springs House.
An imagined nonlinear panorama of the house's interior spaces created in Midjourney and blended together in Dall-E 2.

sketch was limited, but the accumulation of iterative sketches built up a designer's understanding and desire for a project. The challenge was to translate this vision into reality, where most digital tools developed over the last 30 years provide value. However, the early stage of incoherence is of great value to the designer as it represents the inherent contradictions of design.

For programmers in AI, there is a constant effort to produce models that yield more coherent results. However, the most significant value of AI models for designers may be their ability to help navigate ideas without requiring immediate resolution. In a contemporary workflow, designers must construct complex parametric and BIM models that are often fragile and limited in flexibility. We commit early as the exploration of diverse ideas can be too challenging. The images created with generative AI, on the other hand, allow for the investigation of a broader range of options without committing prematurely to specific design directions. These AI models enable the necessary messiness of design, as they do not conform to a geometry-based paradigm. The latent space of the model allows an architecture before geometry.

The Redefinition Of Authorship

The incorporation of AI tools in design poses a challenge to conventional ideas of the solitary genius creator. Since these tools are trained on an imperfect approximation of human knowledge and their interfaces encourage interaction, engaging with them involves a social dimension

vestures, five points of architecture and ai, andrew kudless

Dripping Spring House.
Made in Midjourney v4 with image prompts from Stable Diffusion 1.5.

akin to communicating with a knowledgeable yet eccentric colleague. This interaction is marked by a reciprocal exchange of ideas, as well as frequent misunderstandings and occasional breakthroughs. Consequently, authorship becomes blurred, and instead, we should direct our attention toward the collaborative process of creation and the resulting value of the design.

Design has always been a cooperative effort between a designer, their materials and tools, and the larger social and historical context in which they operate. This collaboration is shaped by a diverse range of living and nonliving systems, including the microbiome within us; our direct collaborators, including clients, engineers, and fabricators; and the broader social, economic, and ecological networks that surround us. The introduction of AI models and tools into this milieu has only further complicated the already intricate sociotechnical interconnections of contemporary design. By providing a vast yet imperfect simulation of collective knowledge, AI has expanded and enhanced the designer's capacity to explore the human imagination and envision new possibilities.

This novel perspective acknowledges the complexity of the design process and rejects the outdated model of the master-builder being fully in control of all aspects of a design. Although our role is to synthesize disparate streams of information and competing criteria, we do this in partnership and solidarity with multiple overlapping and nested communities. The collaboration with generative AI tools adds a new layer to this understanding of shared authorship.

De-Coding Visual Cliches and Verbal Biases: Hybrid Intelligence and Data Justice

Sina Mostafavi and Asma Mehan

The truth is the whole. An empty signifier can, consequently, only emerge if there is a structural impossibility in signification as such, and only if this impossibility can signify itself as an interruption (Subversion, distortion, etcetera) of the structure of the sign (Laclau, 1996b)[1].

The emergence of text prompt to image generation by using various Artificial Intelligence (AI) enabled tools has resulted in a profound impact in the fields of design and architecture. From a technical point of view, it is anticipated that further advancements in Generative Adversarial Networks (GANs), Diffusion Models (DM), and other generative methods will expand the application of AI to 3D and 4D domains. In this context, it is becoming increasingly clear that beyond the technical aspects of these technologies, common sense, critical thinking, and ethics are of utmost importance. Therefore, next to advancing machine intelligence, human consciousness is instrumental in the formation of what we refer to as Hybrid Intelligence[2] that can aid us in surpassing visual clichés and verbal biases when interacting with AI.

In a recently published article titled *The Dark Risk of Large Language Models,* Gary Marcus argues that Humans aren't ready for convincing conversations with AI and the consequences will be serious. He further elaborates that technically AI is moving faster than people predicted before, however on safety, justice, and ethics, it is moving slower.[3] Consequently, it is no wonder that the concept of prompt engineering has become exceedingly crucial in text prompt to image generation. The importance is not just about engineering the prompt in a way that produces the desired results, but it's also about being mindful and conscious of the implicit verbal biases and visual clichés that can be embedded and concealed within these generative processes. If we perceive AI-assisted generative design procedures as a kind of dialogue that yields co-authorship between humans and machines, it's reasonable to regard bias at two levels, given that all conversations involve at least two sides. From a human perspective, expecting the prompt writer to be conscious of all conceivable biases is overly optimistic, if not unattainable. This is particularly true when it comes to geography-specific details and cultures that are unfamiliar to the writer, making it challenging to expect a reflective and mindful approach to what has been co-created. From the machine's viewpoint, the architecture and characteristics of algorithms, as well as the nature of processed and trained data, significantly influence how the generated visual outcome may exhibit discriminatory tendencies towards certain geographies. In the near future, we may observe a growing integration of emotions into AI technology. Nevertheless, Pragya Agarwal points out in her article, *Emotional AI Is No Substitute for Empathy,* that AI tools that make inferences about emotional states are likely to worsen gender and racial prejudices, perpetuate and reinforce existing inequities in society, and further marginalize vulnerable groups.[4]

In this chapter, we share sets of experiments done mostly with the goal of creating blends of material systems and visualization that fuse different geographies and cultures. Beyond the hybrid aesthetic nature of generated images, in the following sections we reflect on how creators can be more cautious about human–machine collective authorship, geo-specificity, and data justice.

Human–Machine Intelligence and Empty Signifier

If signs can be used to tell the truth, they can also be used to lie. [5] (Umberto Eco, 1976)

Umberto Eco's quote emphasizes the idea that signs, which can include language and other forms of communication, are not inherently truthful or false. This is particularly relevant when considering the use of generative AI, which can perpetuate biases and discrimination if it is not designed and trained carefully. With the rise of generative AI, it is even more critical to be aware of the biases and cultural assumptions that can be embedded in the algorithms and data used to train these systems. This can result in unfair and discriminatory outcomes, particularly if the AI systems are not designed and trained to be culturally sensitive.

Persian copper craft tectonics and muqarnas fenestration.

Similarly, Ernesto Laclau is a political theorist and philosopher who developed the concept of the *empty signifier* in his work on discourse theory. Laclau believes that the *empty signifier* is the product of the *exclusionary limit* of a signifying system through distinct effects such as *ambivalence* and *negativity*. No social fullness is achievable except through hegemony, and hegemony is nothing more than the investment, in a partial object, of fullness, which will always evade because it is purely mythical.[6] According to Laclau, an empty signifier is a signifier that has lost its original meaning and becomes disconnected from any specific referent but retains a strong emotional or affective charge.[7] Therefore, an empty signifier arises out of a specific political process in which a particular statement, signifier or practice is transformed into a universality.[8]

Therefore, these effects *introduce an essential ambivalence within the system of differences.*[9] The above-mentioned '*system of differences*' reasserts ideational accounts through the application of the political discourse theory of Laclau and Mouffe.[9] This approach posits ideas in governing discourses to be able to understand how general equivalent demands then become empty signifiers. As Laclau suggests, an empty signifier is a signifier that has lost its original meaning and becomes disconnected from any specific referent but retains a strong emotional or affective charge. In the context of generative AI and text prompt to image generation, this could mean that the AI-generated images and designs become disconnected from their original sources and meaning, but still retain an emotional or affective charge that can influence how they are perceived and received by different audiences.

Generative AI and Geo-Specificity

In the context of data justice, geo-specificity refers to the understanding that data and algorithms can have different implications and impacts depending on the location and context in which they are used. For example, the use of AI-enabled facial recognition technology trained with western data sets in an eastern context may have different consequences for communities of color. The use of predictive policing algorithms may have different implications for low-income neighborhoods than for affluent ones. Therefore, data justice requires a critical examination of the ways in which technology and data are used in specific contexts and the potential implications they have for different communities. This means considering the unique historical, cultural, and social contexts in which data and algorithms are used; we need to design solutions tailored to the specific needs and concerns of the communities they affect.

Challenging the dominance of prevailing narratives brings to light the inherent limitations and biases present in such technologies, emphasizing the need for AI systems to be trained and designed using a varied set of data sources and perspectives. The fact that mass media and the Internet are discriminative toward specific geographies and demographics is clearly apparent in the way text prompt to image generation produces biased visual results affected by negativity, cliches, and exclusions. For instance, the word exploded view combined with different contexts and cultural signifiers may result in generative axonometric exploded views and, in some cases, literally results in an explosion. Therefore, reflecting on how contextual information is processed in generative AI is crucial. In architecture, this urges us to consider both potentials and limitations of these technologies for creativity, innovation, and intellectual ownership in the built environment.

In addition, generative AI systems are limited by their statistical basis and the biases and limitations in the training data, which prevent them from generating definite outcomes. They are better suited for generating content resembling existing patterns and structures rather than creating individual and controlled outputs. For example, a generative AI system trained on a dataset of images of faces might generate new faces that resemble those in the training data. Still, the specific features and attributes of the generated faces will depend on the patterns found in the trained data. This raises interesting questions about how collage, amalgamation, and superimposition of different layers of data and visual information are being utilized in architectural design and documentation through generative AI, resulting in a new aesthetic that is inherently incomplete, eclectic, or hybrid.

Data Justice and the Aesthetics of Incomplete Hybrids

In order to achieve data justice, it is essential to consider the technical aspects of data processing and AI technology and the social, cultural, and political context in which they are used. This involves adopting a human-centered approach to technology development and deployment, considering the diverse perspectives and experiences of all individuals and communities affected by data and technology. In this way, de-coding data justice involves challenging and breaking down how visual cliches and verbal biases can perpetuate systemic inequalities and working towards a future in which data and technology are used fairly and equitably for the benefit of all. We argue that a data justice framework must be grounded in an interdisciplinary approach, drawing from fields such as ethics, human rights, and social justice, and must be informed by a deep understanding of the political, economic, and technological contexts in which data systems operate. When generating images from text prompts, it is essential to consider the concept of incomplete or hybridized socio-cultural signifiers in relation to data justice. This acknowledges that datasets are constantly evolving and never fully complete, absolute, or informed. To address this, we need to move away from the notion of fixed technological solutions and toward a more adaptive and iterative approach. It is also important to recognize that data and technology solutions are always hybrid, combining elements from various contexts and disciplines.

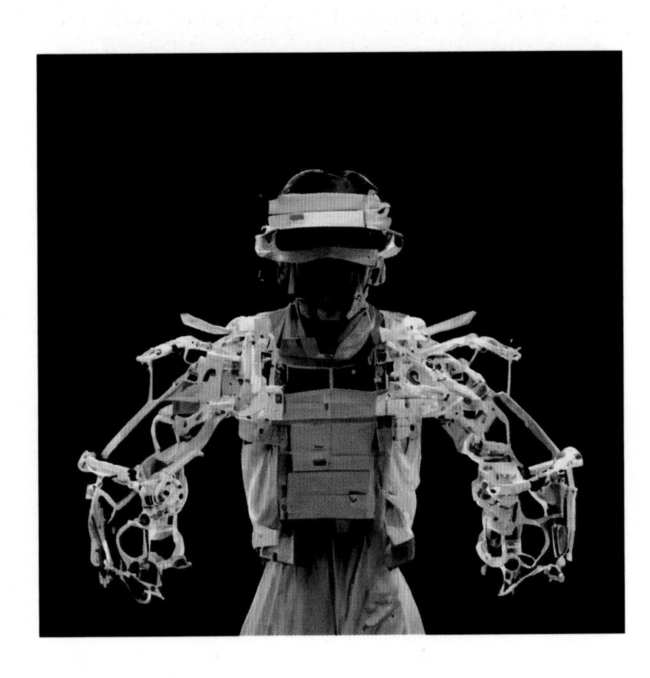

Cyborg or Android.

The explosive growth of AI in various fields, including architecture, has yet to be thoroughly assessed in terms of its impact on the humanistic values that distinguish architecture from technology.[10] Therefore, it is essential to explore new approaches that use AI-generated images to assist architects in designing structures that are more closely aligned with their specific location and context. By utilizing AI-generated data, architects can gain a better understanding of the social, cultural, and environmental factors that influence the design of buildings and structures. Additionally, text prompt to image generation is not only a way to create new forms, but it also offers an opportunity to uncover biases and clichés that reinforce inequalities and preconceptions. Manner, and that the potential benefits of these technologies are realized for all

Where, Why, and How Tetrahedron Voxels Explode?

To investigate the impact of geo-specificity and data justice on generating images from text prompts, we kept three to four keywords constant and changed one word to reflect a particular cultural context. The constant words, including Tetrahedron and Voxels, have a universal meaning related to geometry. However, the interpretation of the word *exploded* can differ depending on the context. This raises the question of why *exploded* can generate axonometric views in certain contexts while being associated with explosions, fire, and aggression in others.

Despite the potential of Generative AI to create diverse and imaginative images, it is crucial to recognize the presence of embedded biases within the technology. These biases are often perpetuated by cultural cliches and stereotypes, which are influenced by mass media and dominant cultural narratives.

Geo-Tectonics and Blended Modalities of Representation

With an emphasis on nations of craft, miniature, perspective, and metamorphosis, these text prompt to image generations investigate how different modalities of representation and materiality can be blended with one another, creating new modes of collective mass authorship. Artists such as Escher and Piranesi have challenged the traditional modes of representation and visualization by purposefully

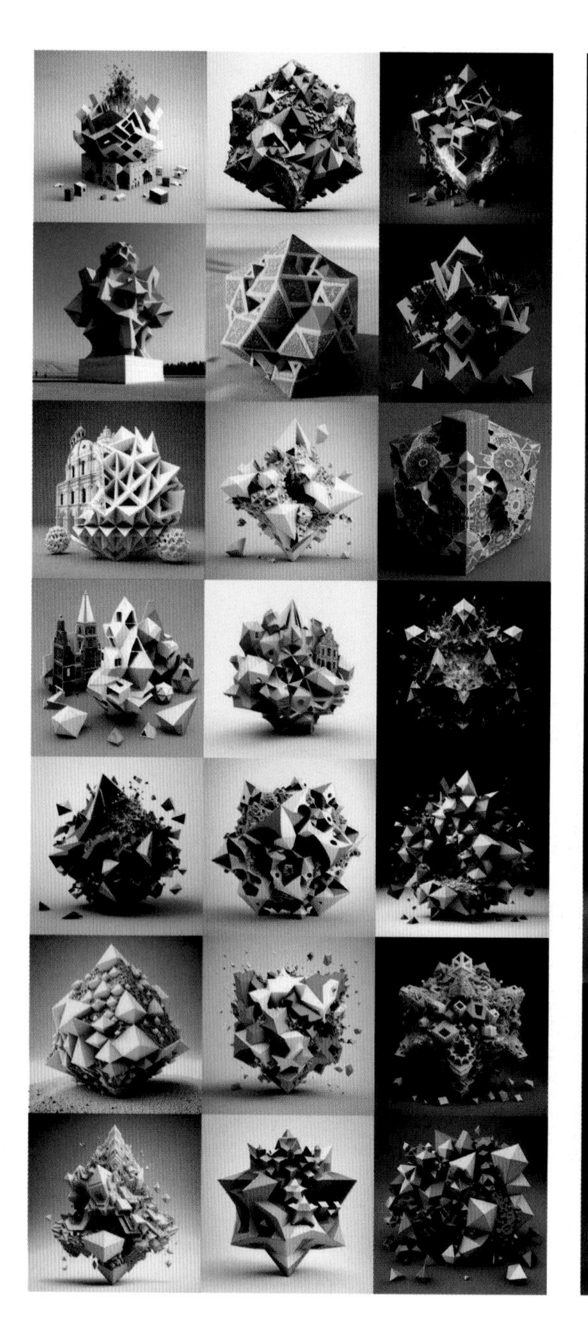

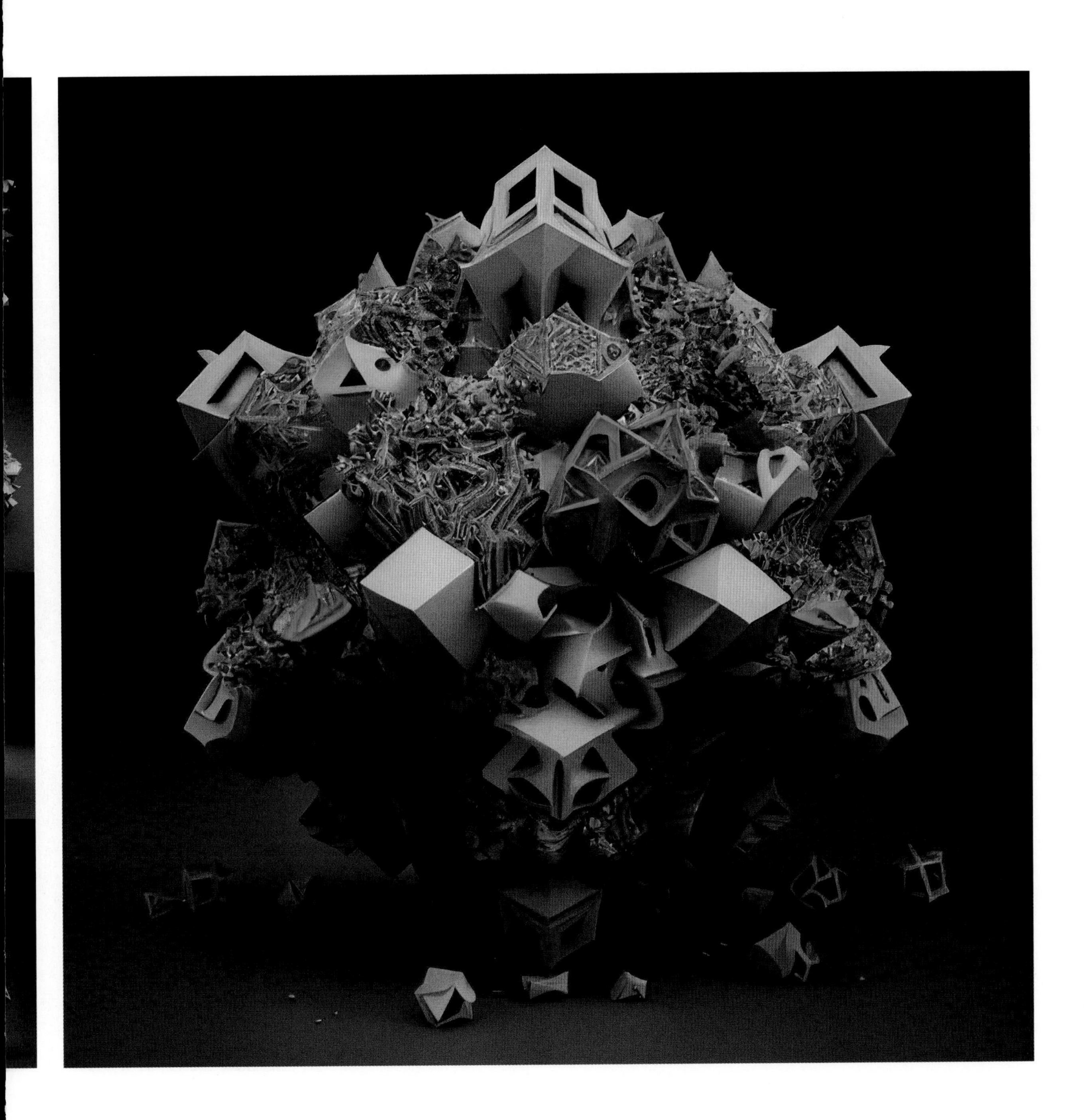

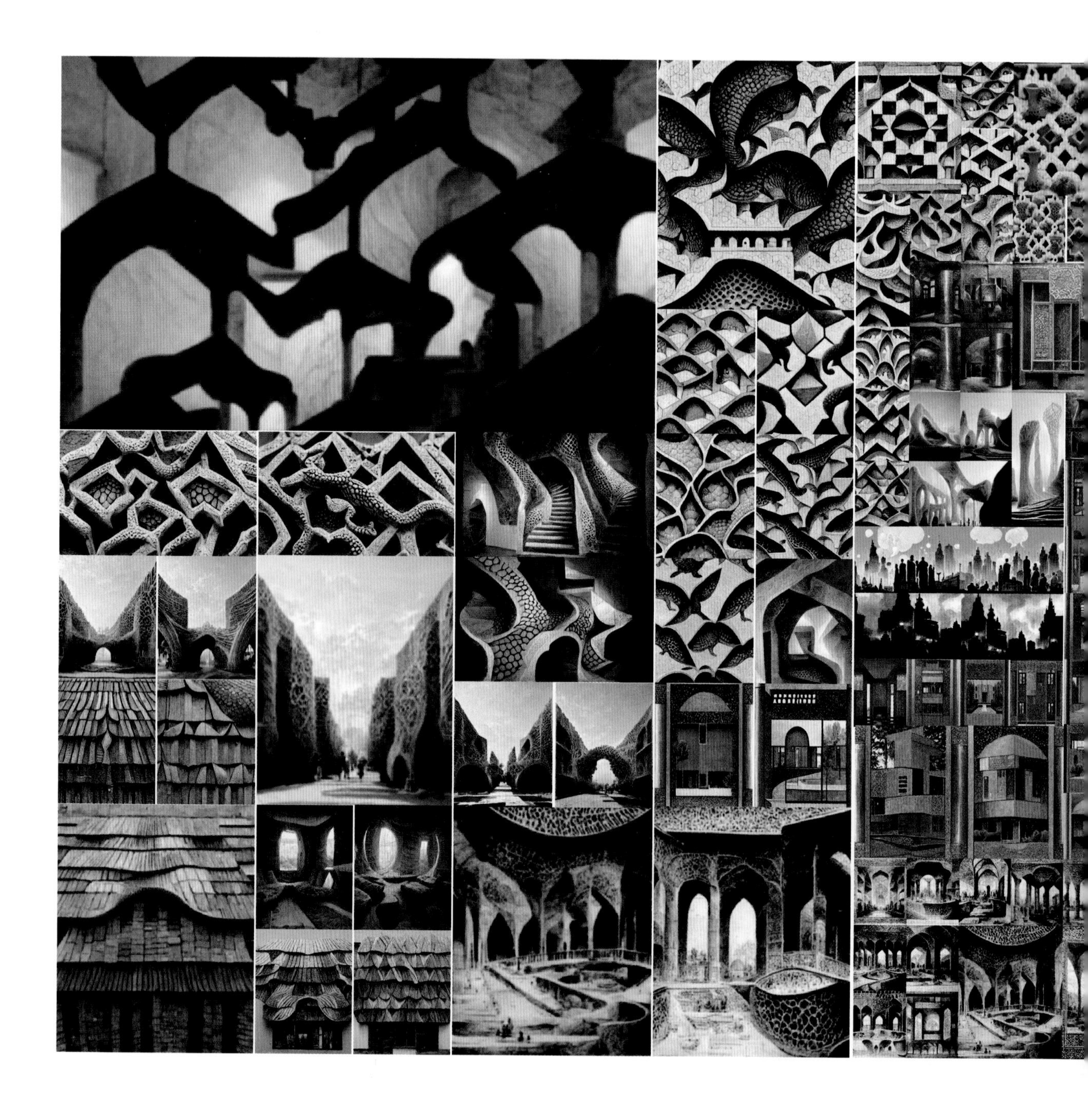

members of society, breaking the principles of perspective and establishing new urbanscapes that incorporate elements of mathematics, metamorphosis, illusion, imagination, and non-Euclidean geometry. In their works, they employed visually impossible geometric shapes that defy the laws of physics, perspectives, materiality, and spatial logic. This approach to craft and representation has had a profound futuristic impact on art, leading to a reimagining of the principles of perspective and the boundaries between 2D and 3D representations.

Persian miniature, on the other hand, offers a different mode of representation that possibly combines both 2D and 3D formats of visualization. This highly detailed and ornate art form emphasizes the use of color, pattern, and intricate design to create an illusion of depth and three-dimensionality within a two-dimensional space. By using various techniques such as overlapping and size manipulation, Persian miniature artists create the illusion of a third dimension, blurring the boundaries between 2D and 3D representations. This technique highlights the potential of combining different modes of visual language to create entirely new forms of art, spatial orders, and geometry, which can challenge the dominant mode of representation that is largely influenced by modern perspective and landscape design.

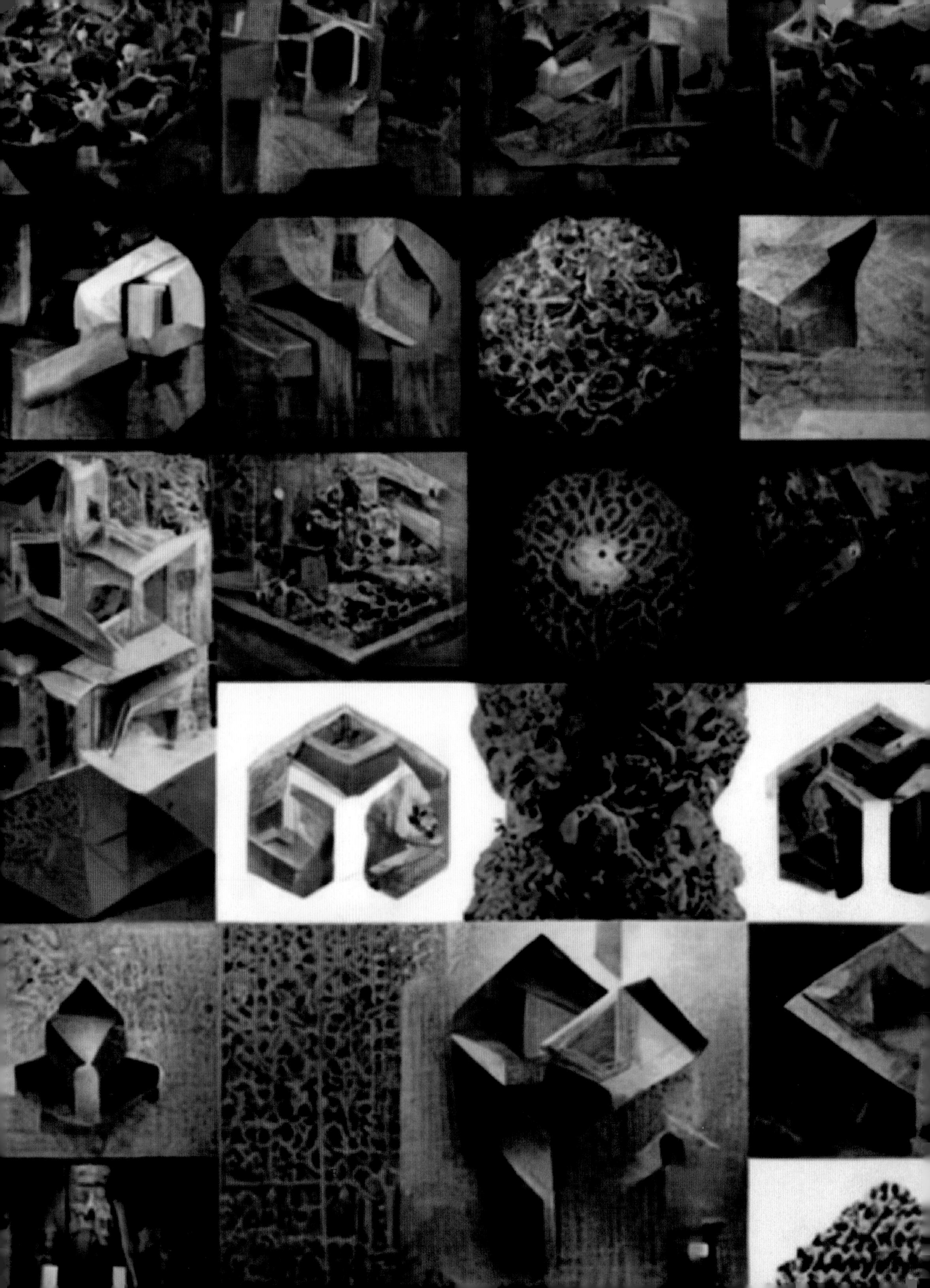

Light Architecture: The Future of Nonhuman Spaces via Circular Thinking for Urban Lifestyle

Rasa Navasaityte

Design Light architecture is a concept that explores the potential of using light to reduce the carbon footprint of buildings and improve urban spaces. This approach focuses on the role of light in shaping buildings and cities, and the ways in which it can be used to create more sustainable and livable environments. The lighting of buildings has always played a determining role in the form of building typologies. The depth of natural lighting, and the standards derived from it, long limited the dimensions of building typologies to the depth of two mirrored living rooms. The endless dimensions of 20th-century warehouses and similar buildings were unthinkable without artificial lighting, expanding human sight for work, play, and navigation. Artificial intelligence (AI) will add a new complex play of forms by expanding the spectra of light. In the orchestration of complex networks, light becomes the flow of information and energy: photo and synthesis.

Planting UV Light examines the potential for indoor urban farming to shape architecture.

Design processes in architecture and urban design are constantly evolving, driven by new technologies and changing societal needs. In the field of architecture, the increasing focus on reducing carbon footprint has prompted designers to find new ways of designing spaces that are both environmentally friendly and appealing to humans. This has led to the exploration of new urban perspectives, where non-human spaces are integrated into building designs, creating a symbiotic relationship between the built environment and nature. One of such an approach is the use of AI to create new urban interiors that are designed and lighted by synthetic

By collaborating with AI, the project imagines the large communal spaces and architectural benefits that can be achieved with indoor greeneries.

vestures, light architecture, rasa navasaityte

Synthetic Nature and Synthetic Light knits nature into the building.

nature and synthetic light. Through the use of AI, city spaces can be transformed into livable places that incorporate light, materials, and plants, creating a new form of synthetic symbiosis between nature and the built environment. One example of this is the project *Planting with UV Light*. This project explores the synergies between urban indoor farming and architectural design.

With recent advances in technology such as machine vision, automated seeding, and low-energy UV lighting, the compression of agriculture into vertical indoor farming has become a reality. The project envisions large communal spaces and other architectural synergies within the thick interior of indoor greeneries. The UV lights needed for vertical planting are also seen as opportunities to create new urban interiors and stimulate the integration of nonhuman spaces into the city.

Another project, *Trees in Daylight,* explores the potential of bringing greenery and density closer together by using AI to create a symbiotic relationship between buildings and mature trees. With AI's hyper-compositional qualities, the project envisions a porous building that incorporates mature trees into a vertical forest, thereby creating new densification opportunities for cities. The vertical arrangement of trees is a symbiotic part of the building, where trees are an integral part of the building's design, connecting and interlacing technical, industrial, nonhuman streetscapes, food production, urban gardening, pedestrian floats, and more. The project *Synthetic Nature and Synthetic Light* weaves aquaponics and micro gardens into urban interiors, designed and lightened by AI. The weaving with light creates new urban spaces that are more liveable and healthy for people. The goal is to turn city spaces into liveable places by adding light and lighting materials, elements, and plants. This project envisions a synthetic symbiosis between existing urban spaces and nature, renovating, recreating, and reactivating city spaces.

The *Austin City Streetscapes* project explores the potential for increasing density and incorporating more tree areas into the city's streetscapes. By re-imagining existing concrete pavements and interconnecting them with new massing, bridges, and pathways, the goal is to create more urban landscapes that improve human flow and overall livability in the city. When cars become electric and driverless, our perception of nonhuman spaces in cities is set to change. These areas, once dusty, loud, and polluting, will become quiet, calm, large, and wide. This opens up new opportunities for affordable ground conditions, which can be used for walkability, plazas, interaction, food production, and more. This new light for a different future has the potential to become a new ground for human areas, with public squares, food production, and parks all being incorporated into urban life.

The circular benefits of these projects enable us to explore the potential to create closer relationships between nonhuman spaces and urban lifestyles. Here, people create their own spaces within nonhuman areas that provide food, homes, and shared areas for interaction. These areas should not be designed purely for financial investment, but

Using light and lighting materials, elements, and plants to create livable city spaces.

vestures, light architecture, rasa navasaityte

The yellow light used in this project signals a synthetic symbiosis that weaves into existing urban spaces, creating a potential for nature and people to coexist.

vestures, light architecture, rasa navasaityte

should also involve human creative involvement in order to enhance the value of the city and create new city centers. The potential for the integration of nonhuman spaces into the urban fabric is immense. New possibilities are opening up for the integration of vertical indoor farming, urban gardens, and food production into the design of cities. This not only has the potential to reduce carbon footprints and improve sustainability, but it also provides opportunities for socializing, interaction, and the enjoyment of new potentials within the urban environment.

Trees in Daylight focuses on the role of light in bringing nature and density closer together.

By re-imagining existing concrete pavements and interconnecting them with new massing, bridges, and pathways, the goal is to create more urban landscapes that improve human flow and overall livability in the city.

Driftwood Rock

Igor Pantic

{}

Timber Beach Houses

Driftwood Rock series explores a blend between the expressive nature of driftwood and geometric architectural elements. What initially started as structures placed on the ground was curated towards elevated structures with exaggerated cantilevers. The upper portions of these structures fully express the ornamental and textural qualities of driftwood, while their base is often neither rock nor driftwood, but rather an unrecognizable blend between the two.

vestures, driftwood rock, igor pantic

C

vestures, driftwood rock, igor pantic

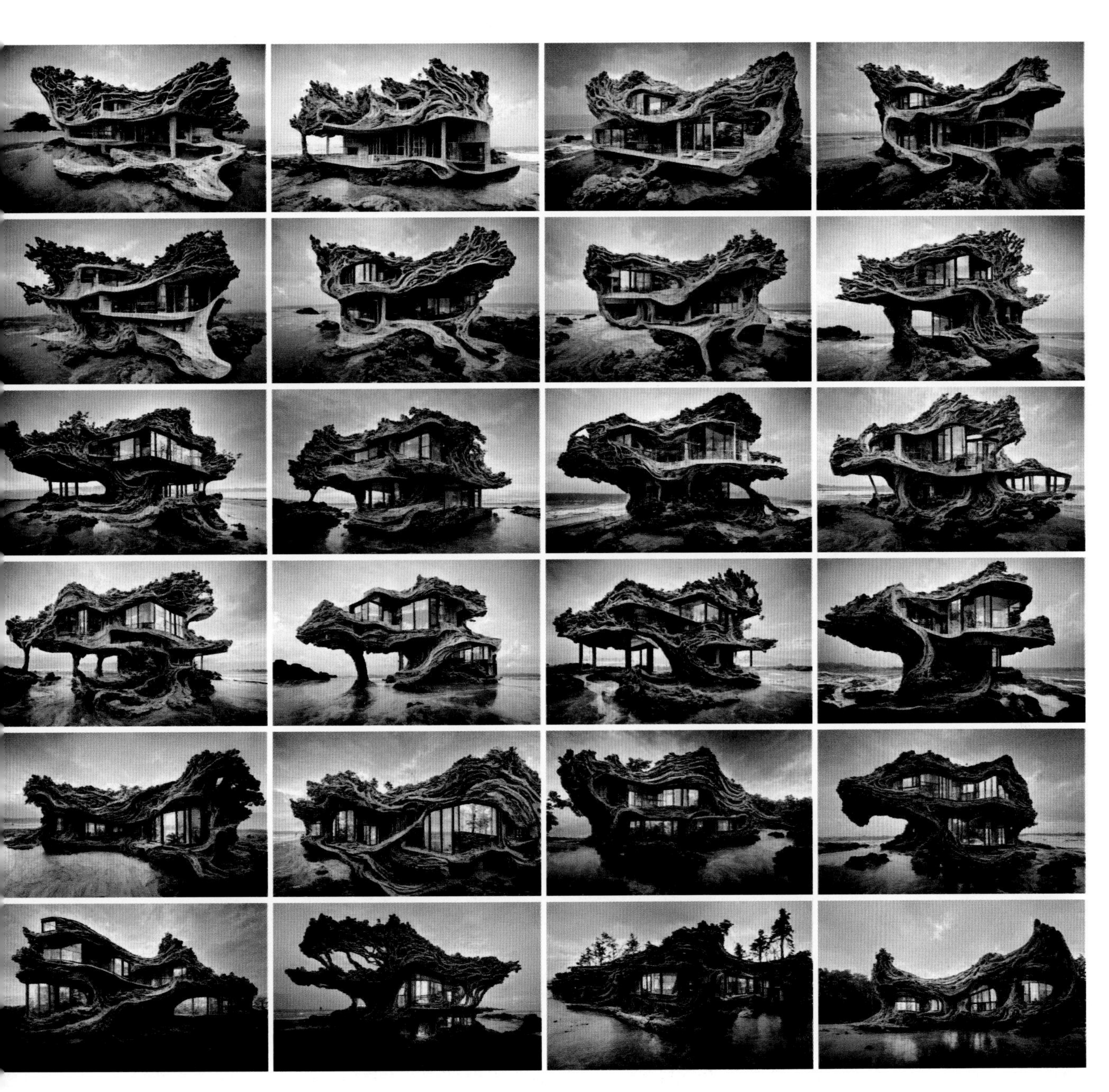

C

Timber Forest Houses

Darkwood Dub series plays with the dialogue between tectonic geometries and the inherent organic features of raw timber. Together with this, the recursive ornamentation of the facades creates structures that at the same time appear ancient and futuristic, mechanic and organic.

vestures, driftwood rock, igor pantic

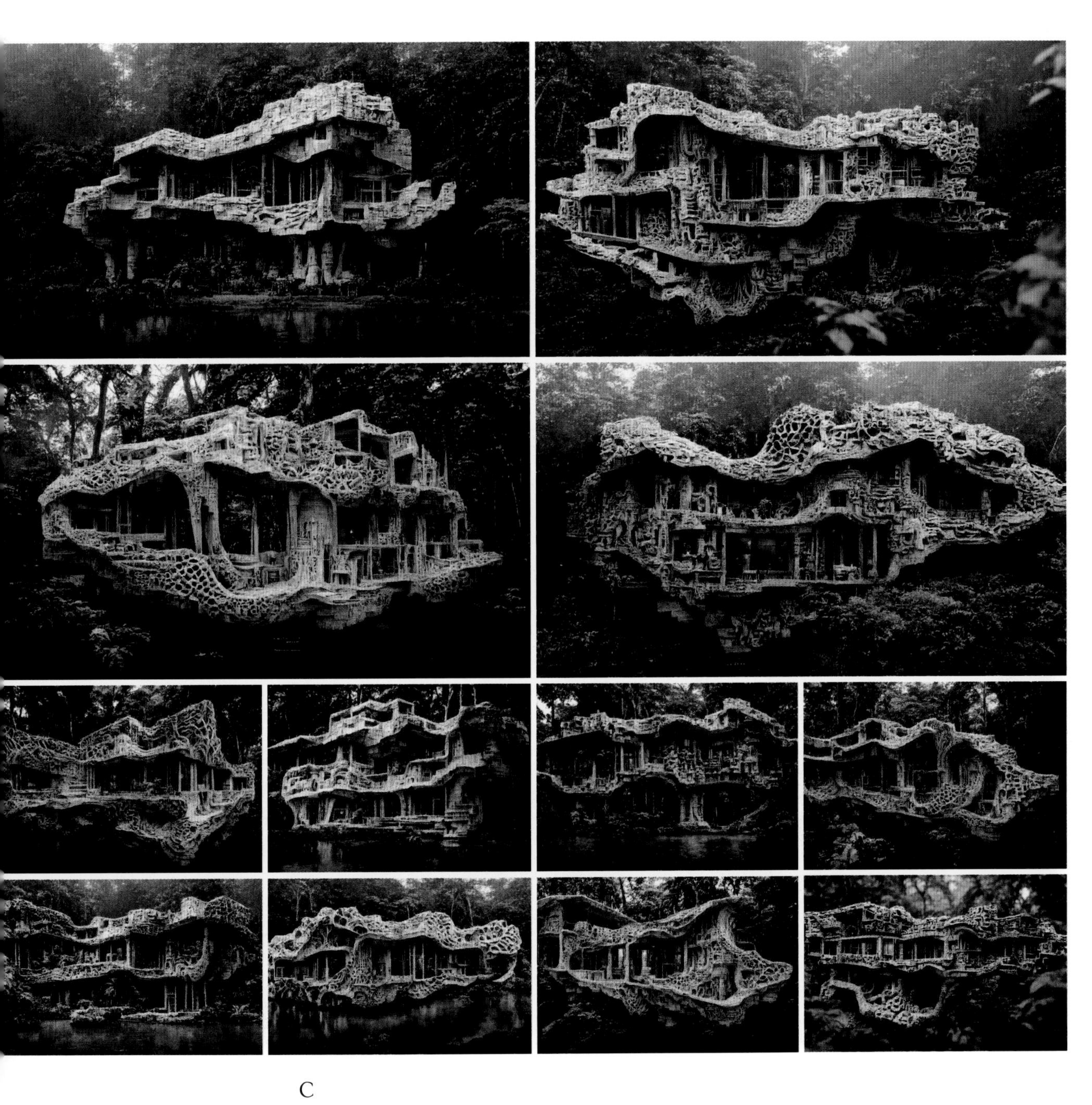

C

Material Driven Forms

Latent Materiality explores the formal and textural qualities of timber and stone at the scale of sculptures and deep surfaces while taking into consideration the properties of each material and the method of formation. Instructions such as recursion, delamination, erosion, voxelization, and carving had the purpose of replicating algorithmic design methods.

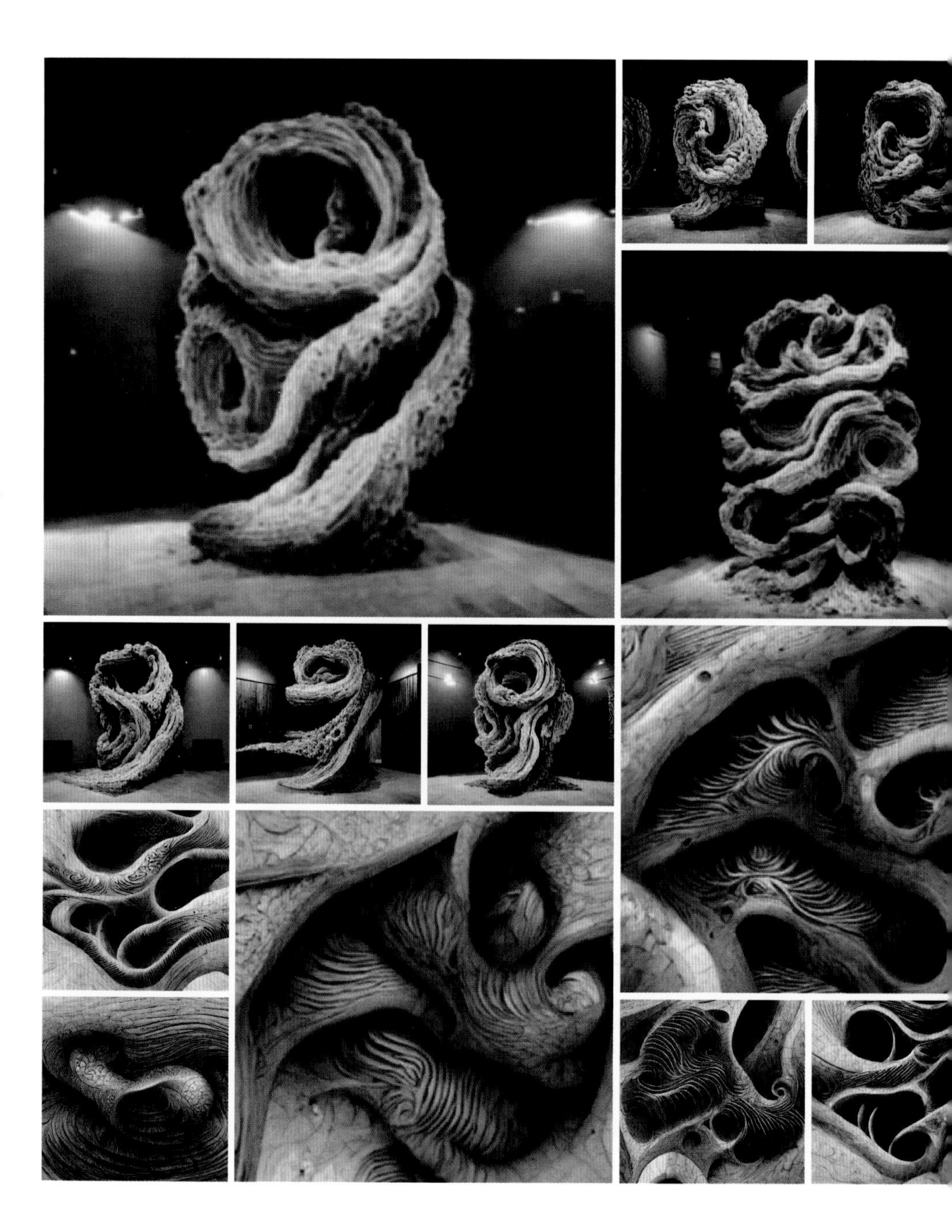

Carved, laminated, and steam-bent timber structures imagined out of driftwood and hardwood. Some of the results appear multi-scalar and could be read as a close-up detail, openings on a surface, or pockets of space.

vestures, driftwood rock, igor pantic

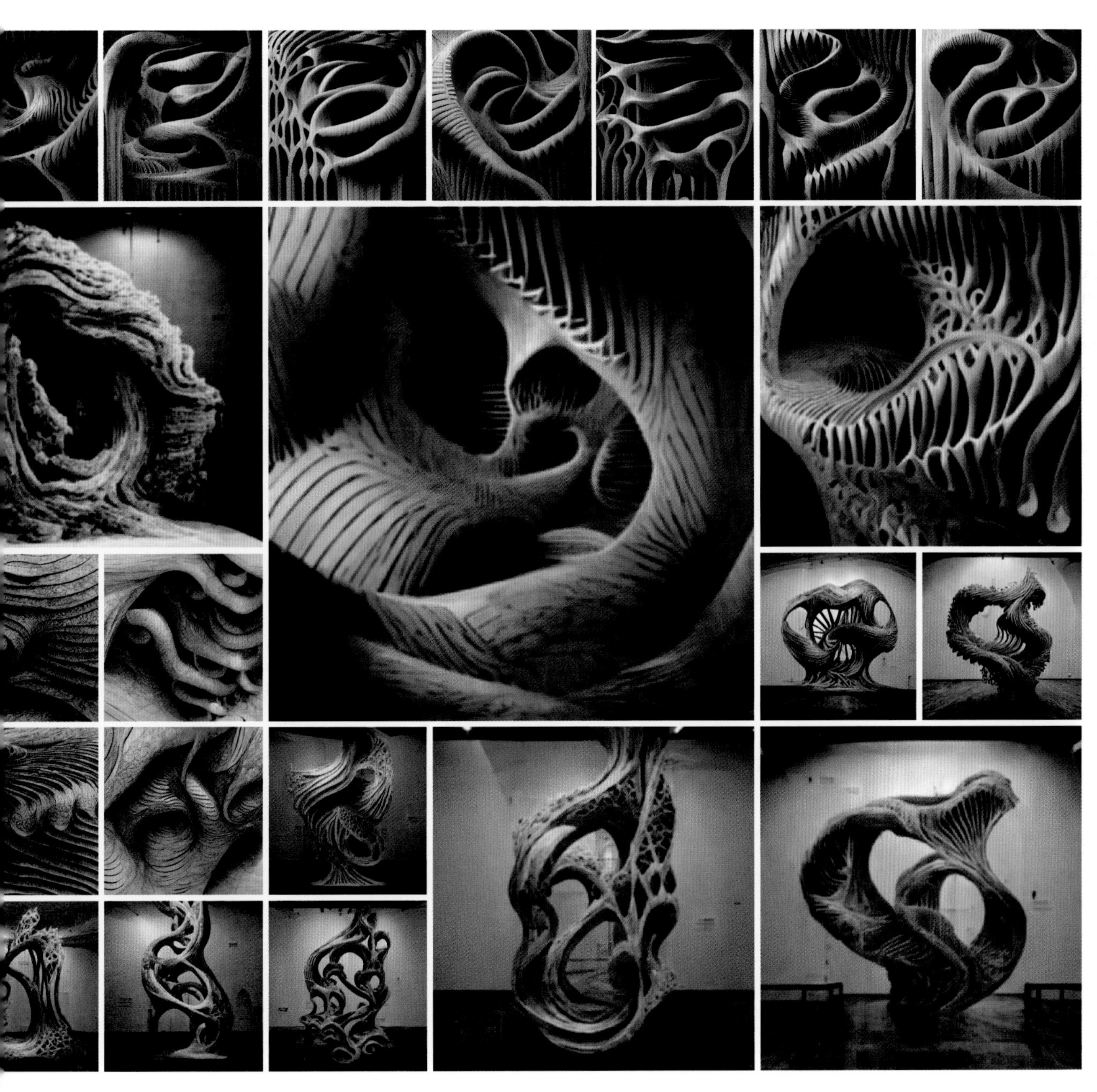

C

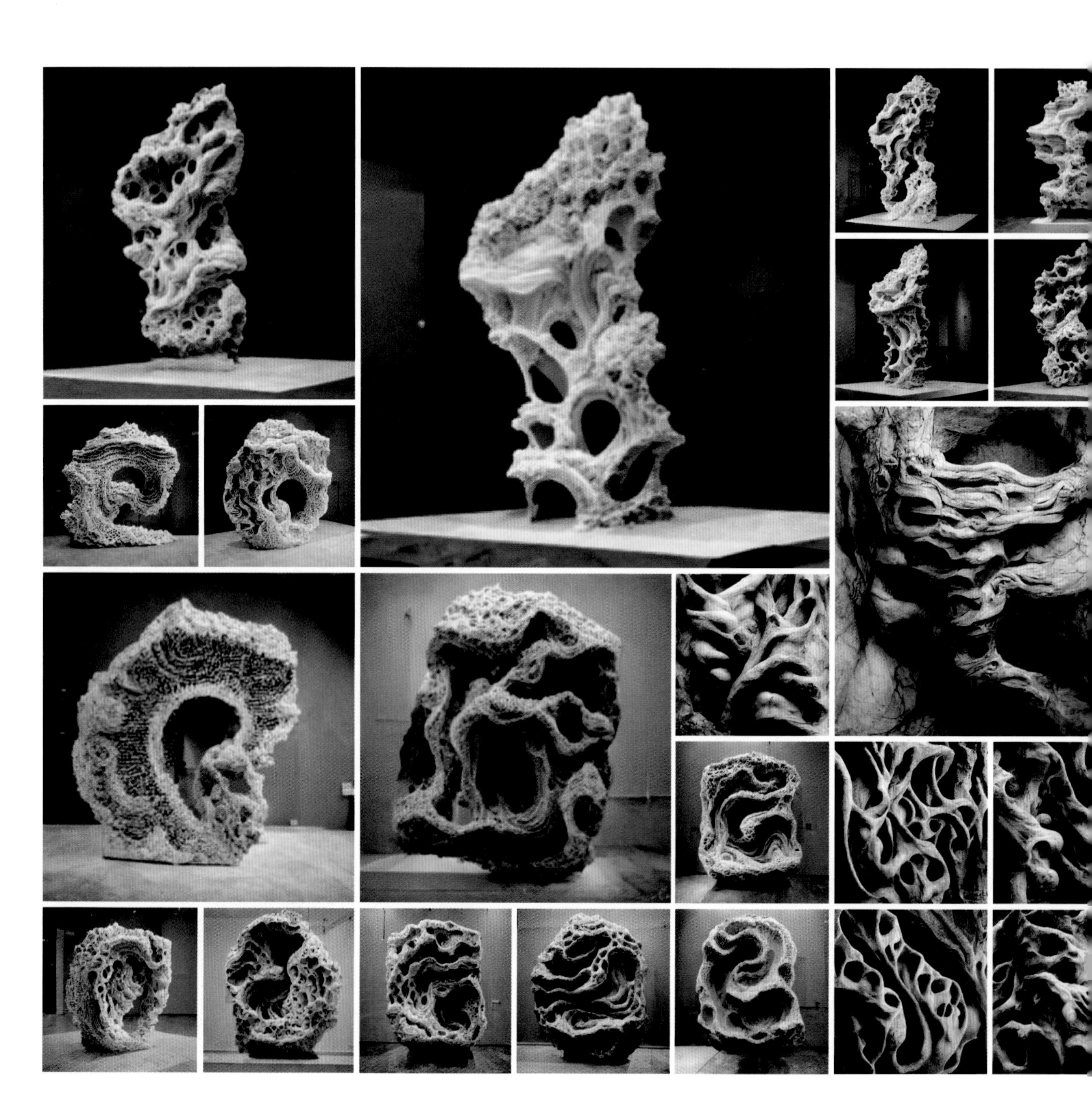

Stone sculptures exploring combinations of formation methods (erosion, cutting, carving) with different types of stone (scholar rock, sandstone, marble etc.)

vestures, driftwood rock, igor pantic

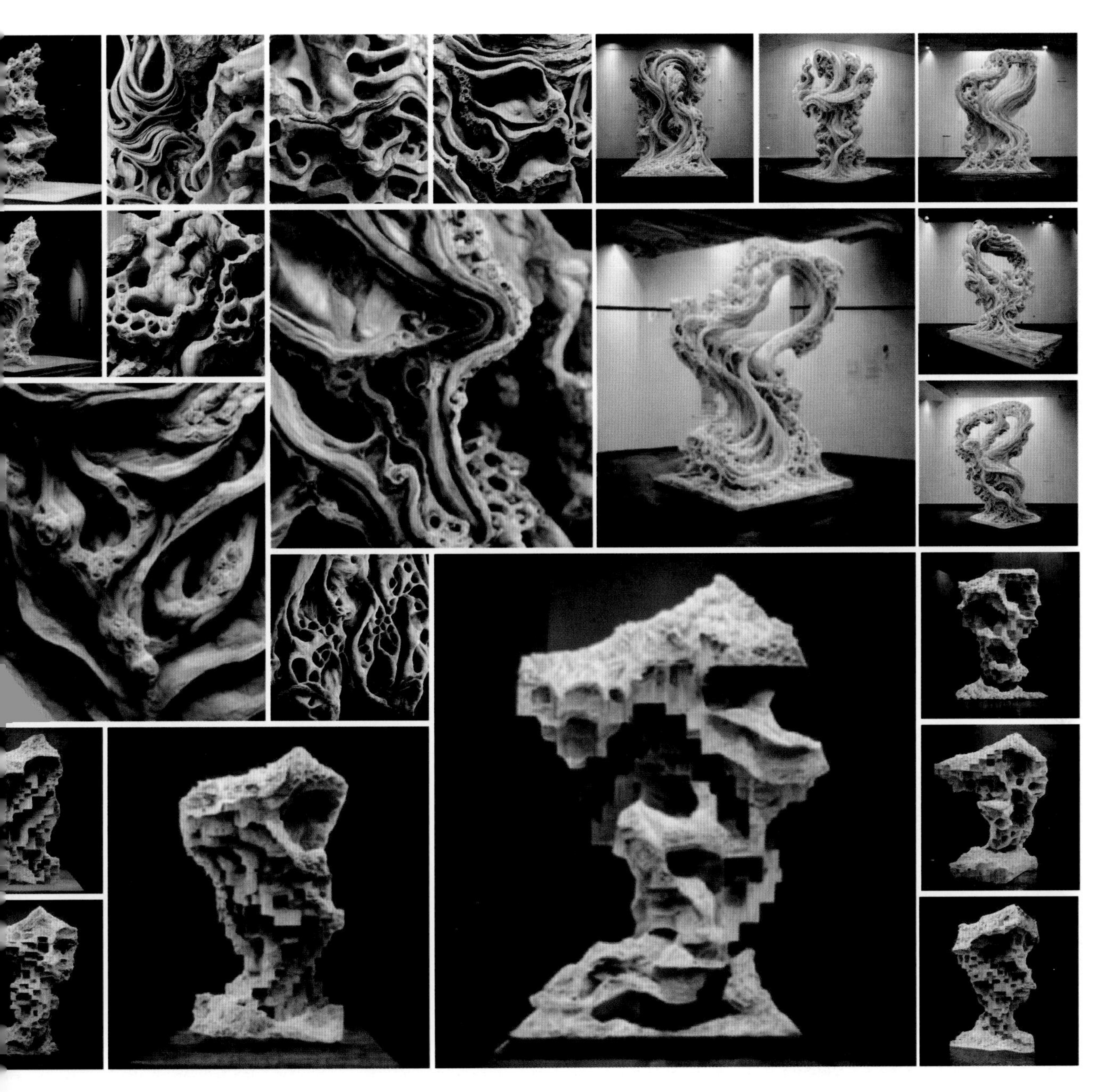

C

Digital Spoliare

Dustin White

{}

Memories, experiences, thoughts, and questions all play a significant role in the process of generating Diffusion models. We can now combine our dreams, experiences, random thoughts, and technical prowess with a human touch to enrich our architectural design process. With this fusion, we can make our aspirations and experiences a reality, bringing to life designs we never believed possible. The future of material practice and manufacturing has made it feasible to achieve aesthetics of complexity with simplicity, providing a unique opportunity to delve into the history of architecture and generate new possibilities for form and ornament. In my personal experience making thousands of AI-generated images and sketches, I, like many others, have been exploring a varied spectrum of topics, but a constant thread of language concentrating on pieces, components, materials, and compositions directly tied to the idea of "spolia" has emerged. Spolia can be defined, in archaeology and art history, the phrase refers to the reusing of architectural elements or inlaying the memory of earlier projects onto new ones for decorative effect.

The relationship between diffusion models produced by Midjourney and the concept of spolia revolves around the idea of reuse and preservation of memory. AI diffusion models produced by Midjourney aim to simulate and understand the spread of AI technologies in complex systems, such as organizational networks and social networks. The models consider various factors that drive or hinder the diffusion of AI, including network structure, technology type, and the characteristics of organizations or individuals. AI diffusion models can preserve memories of previous styles through memory networks, style transfer techniques, and

vestures, digital spoliare, dustin white

/imagine: An enchanted contemporary vault house, featuring a unique and flowing cloth mesh structure. This visionary design blends the beauty of Japanese landscape with a futuristic twist, incorporating a stunning redshift effect. The dramatic lighting and pink bloom aesthetic add to the epic and hyperrealistic feel of the space. This house is a true celebration of the intersection of art and technology, as every detail has been rendered with hyperrealism and photorealism.

training on large datasets. The process is similar to the idea of spolia, where fragments from previous structures are reused in new building projects to preserve memories of the past. This historical tradition serves to reduce the extraction of new, virgin materials. Why go to the effort of sourcing, cutting, and transporting materials when they are available locally?

In addition, both AI diffusion models and spolia also serve to preserve the memory of previous constructions or technologies. AI diffusion models capture the interactions and spread of AI technologies, thereby preserving the memory of technological innovations and their impact. Similarly, spolia has a unique ability to reinvent itself to meet evolving communal needs, a bricolage of elements, ideas, and memories recomposed and preserved into a new structure. This process became the focus of the early rules sets I was developing using expansive design practice to analyze both Gothic and Art Nouveau elements. The process of collaborating with the tool to distort, split, and rearrange elements to fit into a given space, but it's important to note that the essential body is apparent, and to some extent, their relationship to one another is maintained.

The following are examples of rules that were created to establish the prompt structure: (1) stylizing as opposed to realistic representation, (2) schematic characterization by accentuating certain features, (3) dislocating split details.

In closing, the relationship between AI diffusion models produced by Midjourney and the concept of spolia or digital spolia highlights the importance of reuse and reduction of extraction. Both concepts offer valuable perspectives and solutions for the development and preservation of memory, whether it is in the form of technological innovations or building materials. By reducing the extraction of new materials and

/imagine: Irrational alterations of combined art nouveau and gothic architectural exaggerated features into new representations.

preserving the memory of previous constructions or technologies, both AI diffusion models and spolia contribute to a more sustainable and efficient future.

/imagine: Analyze defining elements of Gothic architecture, such as the Pointed Arch, Rib Vault, Flying Buttresses, and elaborate Tracery, analyze these elements and use them as the basis, through a process of distorting, splitting, and rearranging these elements to fit into a given space, stylizing the elements instead of creating realistic representations, transforming details into new Gothic representations. – v 2.

/imagine: The art of architecture meets the science of photography in the stunning chartreuse and white porcelain building. Admire the intricate details captured by photographers using Hasselblad cameras. Appreciate the professional lighting and post-processing in Lightroom, which bring the building's beauty to life with incredible sharpness and detail.

/imagine: A two story house inspired by Marcel Duchamp : the house is a ghost house : Ghost is a metaphor a phantasmagoria.

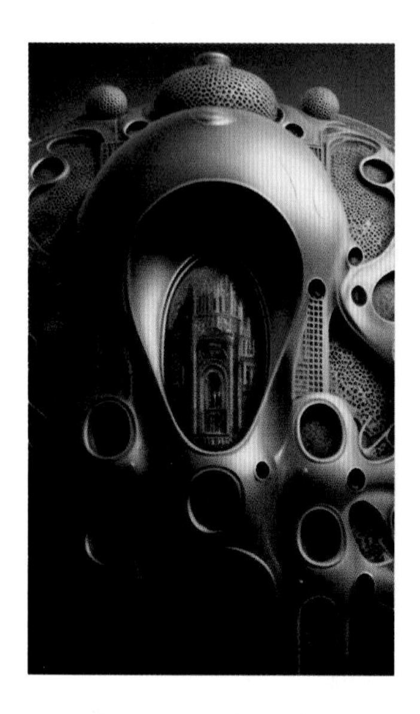

/imagine: The architectural facade of the building is made of blobby, ornate stainless steel:: The intricate details and matte finish are awe-inspiring:: A photographer captures the beauty through different angles and perspectives with the goal of editing and sharing the photos with the world.

/imagine: A sofa made of cojiform parts:: cojiform parts organized in a radial configuration:: constructed from ceramic and deep blue textile.

imagine: "Pop Art Playfulness," refers to the concept of reinterpreting Art Nouveau architectural details in a playful, pop art style. This transformation takes the typically elegant and sophisticated style of Art Nouveau and adds a sense of fun and humor to it through the use of bright colors and exaggerated forms.

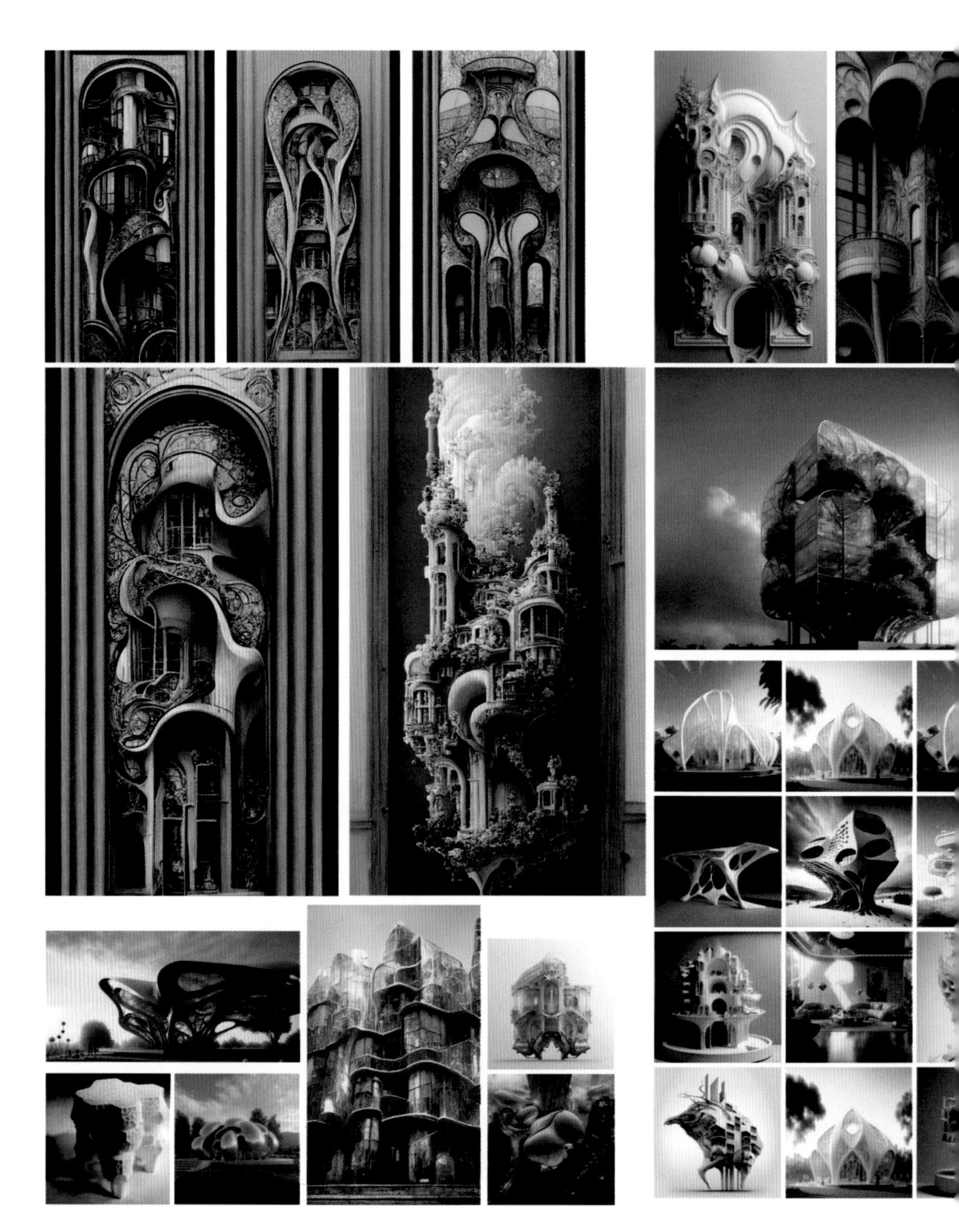

/imagine: Step into a mesmerizing world of light and technology as you immerse yourself in the interactive lighting installation, This progressive exhibit combines innovative materiality technology and computational design, inspired by the works of world-renowned artist, Alexander McQueen. Experience the interplay of light and movement as you navigate through an ethereal landscape, pushing the boundaries of conventional design and igniting your senses.

The concept of estrangement in architecture exists within the interstice of the recognizable and the defamiliarized, the known and the enigmatic, and the ordinary and the evocative. It is a complex and nuanced phenomenon that eludes easy comprehension, necessitating a deep exploration of the language of architecture – its forms, materials, and spaces. The projects in this section live happily in this territory of interrogation. At its core, estrangement serves as a means to challenge assumptions about the world we inhabit. It involves unraveling our perception of architecture (built and unbuilt), pushing beyond the boundaries of our habitual understanding, and allowing us to see the world in a new light.

This idea is not new and has been explored by philosophers and thinkers throughout history. G. W. F. Hegel's examination of estrangement as a tool of self-reflection,[1] Karl Marx's alienation of the working force,[2] Jacques Lacan's charting of the unsettling territory between the conscious and the unconscious,[3] and Theodore Adorno's musings about the contradiction and alienation of modern society.[4] Essential for interrogating the projects presented in this book are the concepts presented by thinkers and artists, such as Sigmund Freud's analysis of the uncanny,[5] Victor Shlovsky's Ostranenie,[6] and Bertolt Brecht's Verfremdungseffekt.[7] In the realm of visual arts, Mario Klingemann, Feileacan McCormick (a trained architect), and Sophia Crespo's works offer compelling examples of estrangement. Klingemann's use of machine learning algorithms to create surrealistic images challenges conventional ideas about the role of technology in artistic production.[8] On the other hand, McCormick's and Crespo's AI-generated menageries of strange animals, such as the "Neural Zoo" series, present an otherworldly aesthetic that defies easy classification.[9] However, estrangement is not limited to artistic expression. It can also serve as a powerful force for social and political change by challenging our presumptions about the world and inspiring us to imagine new possibilities, as, for example, described by Katja Hogenboom.[10] In this sense, it is a tool for subversion, a means of disrupting dominant narratives that shape our lives and opening up new spaces for creativity and innovation.

Ultimately, estrangement is a multifaceted concept that resists easy definition or categorization. By exploring its diverse manifestations and examining its potential as a tool for artistic and social change, we can begin to understand its significance in shaping our perception of the world around us.

///// ESTRANGEMENTS /////

Imago Mundi Imaginibus Mundi

Cesare Battelli

The recent introduction in the international scene of Artificial Intelligence Laboratories reopens a reflection on the role of imagination applied to machines and the paradox that it is precisely from the most advanced technology that we see ourselves retracting the meaning of imagination- imaginary.
From an onto-gnoseological point of view, imagination (not to be understood in this case as the production of unreal worlds) has a long history that spans much of the history of art and philosophy from ancient times to the moment when in the 18th century, at the beginning of post-Cartesian modernity, it ends being the gnoseological foundation of all modern culture. The imagination, specifically not only the transitive imagination, but also the active one (from Synesius to Marsilius Ficino), during the 1400s, was considered the magical foundation of connection between opposite worlds, between intelligible and sensible, macro and microcosm. For Ficino, for example, imagination converts the language of the senses into a fantastic language, so that reason can grasp and understand the phantoms deposited in memory, while active imagination plays an intermediary role between the two shores (sensible and intelligible), *activating* a world of its own. The image as the union of the particular and the universal is configured first and foremost as a symbol, a figure that is both concrete and universal, abstract and figurative at the same time.

These assumptions, elaborated later in Giordano Bruno's ontology, which will arrive at a concept of synesthetic imagination and infinite fantasy, (existence is for Bruno a permanent possibility, a continuous imaginative threshold), are not only anticipators of the 17th-century genre of imaginary art and architecture (Blake–Piranesi), a genre that incorporates within itself the modern Neoplatonist fantastikon in the manner of the simulacrum, but also the assumptions through which the imaginative mechanisms of artificial intelligence (AI) are to be explored.

Based on a gigantic database through a system of algorithms, intelligent machines are capable of imagining, and they do so by seeking a common denominator –unimaginable to the human mind except through lengthy interdisciplinary research– among the various verbal inputs entered into the prompt. Just as imagination is a bridge between opposite worlds, similarly machines create common denominators or mediations between disparate things. The connections that AI labs create between the various components in the text are not simply a collage of intentions, where each fragment is still affected by its origins, but a fusion that is not only the way in which algorithms lose the finite and distinct appearance of any of the inputs, continually developing or blurring them together, but also the atmosphere produced by a lab like Midjourney that can make the relationship between landscape and architecture palpable and architectural.

However, this is not an unrealizable or completely imaginary architecture, but one that is phenomenologically interpretable and thus ultimately realizable. Each image produced by the machines is unique and unrepeatable just as the point of view is unique in that –it no longer deals

with the observer in the finite -infinite relationship and vice versa as in the usual representations in perspective or axonometry- unique and unrepeatable becomes the sequence of inputs that offers in the infinite variations architectural objects that are different but genetically the same.

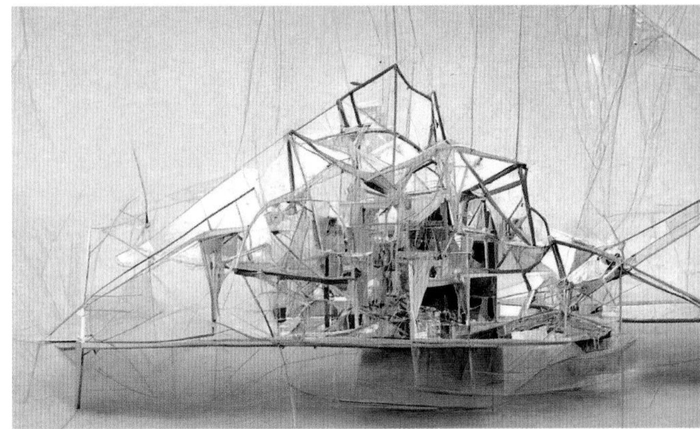

C

Portable Theatrum Mundi

Hybrids

Like the imagination, hybrids also works on the threshold of the possible. AI is an excellent mechanism of hybridization not only between physical things but also with conceptual traces and philosophical quotations, paintings, and architectures. Hybrids are monsters in that they subvert taxonomies and are creators of new categories, through which the dimensions of custom and familiarity are violated. The ancient world was indeed populated with creatures on the borderline between human and animal (think of the hybrids before the Great Flood according to myth), but there were also hybrids that fused into one body elements belonging to different animal species, male and female natures, the androgynous, fatal monstrum, prodigy, and mortal and immortal components. Added to this category are those human beings (Kafka) subject to a process of metamorphosis, locked up in

C

{}

Architectural Fire

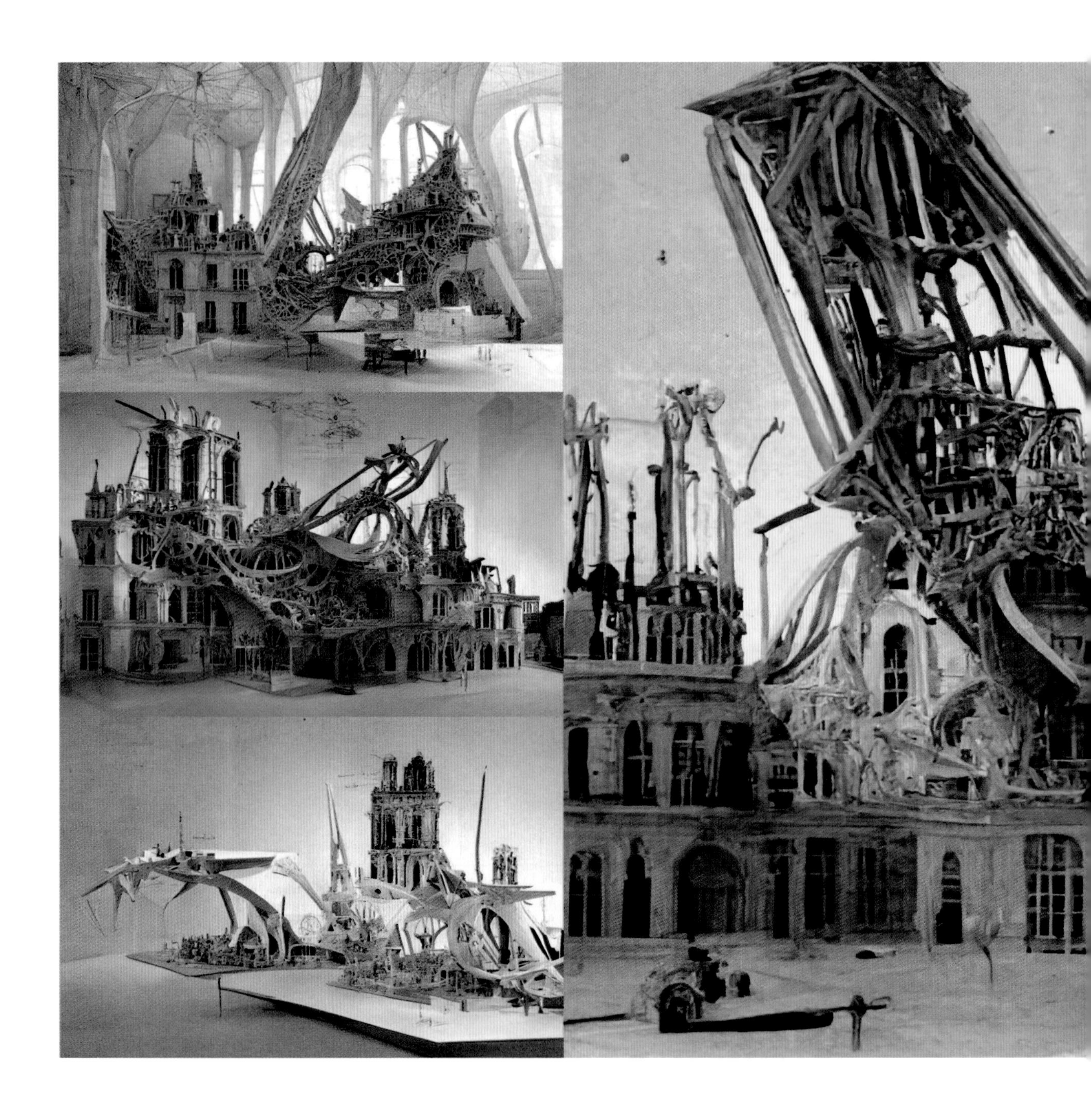

C

"Architecture refers to the design of a building or structure. It encompasses the layout and functionality of a structure, as well as its aesthetic appeal."

an animal body but endowed with human thinking faculties, mixed creatures in the psychosomatic complex of their being. Monsters are in fact the representation of the limit. They are by definition on the border between two natures, liminal beings who, as such, cannot but be placed geographically on the edge of human and civilized society.

Machines Producing Machines

One of my favorite aspects of the hybridization process carried out with AI is to bring together traces (simplified images and concepts) of both magical modernity (especially in terms of pictorial references) and scientific modernity, characterized particularly by the human–machine relationship. The intention is to elaborate architectural machinery, indefinite and imprecise, halfway between solid bodies and all the opposite – between open spaces mixed with closed ones, urban visions, and interior spaces. The relationship between machine and human is by no means new, belonging to a debate as old as it is distant in time since the beginnings of the first steam engines and how the machine itself has begun, in more recent times, to enter fully into intellectual debate and artistic production. From about 1850 to 1925, a large number of artists, especially writers, imagined the workings of history, the interactions between the sexes, and the relation of man to a higher instance translated into a simple mechanics. Freud himself had defined the psyche as an apparatus. Exemplary is the case of Marcel Duchamp`s Grand Verre. This Work can be read in a multiplicity of ways: as a closed cycle composed of an action produced in the upper zone, what Duchamp defines as the bride or milky way but also an inscription (prompt?), and another lower one defined as the bachelor zone. In principle, the machine works as follows: in the upper half of the Grand Verre, there is an amorphous composition, the one belonging to the gestalt world of the fourth dimension. Joined to this is the female skeleton of a bride, which represents a possible projection of this four-dimensional form onto the three-dimensional plane. By means of one of its elements it projects to the area below constituting its nexus. In the lower half, we are in a perspective world, that of the third dimension, that is, the world of celibate machines. In the form of gas, these bachelors form the "graveyard of uniforms and liveries." Mounted on a sled, their movements are stereotyped, while the energy they cause is channeled, through funnels, into the chopping and other gears without a precise understanding of their operation.

In the postwar period, French writer Michel Carrouges, identified the myth of the celibate machines not only in Duchamp's work but also in Franz Kafka's Penal Colony. Indeed, in both machines, there is an upper zone with an inscription that, through a pantograph, transmits a message to a lower zone; in Kafka's tale, it is the recording of the sentence in the condemned man's back by means of the harrow.

3D Diffusion or 3D Disfiguration?

Immanuel Koh

...her face filled eight boxes ... in one, for instance, her eyes were laughing cruelly, but in the next they were filled with sadness. (pp117-118)

The passage above is a quote from the novel "Klara and the Sun" written by Kazuo Ishiguro and published in 2021. Here, Ishiguro, winner of the 2017 Nobel Prize in Literature, masterfully narrates from the perspective of the main character Klara (one of those robotic humanoids called Artificial Friends or AFs in the book) during a trip to a waterfall with its owner's mother. More specifically, Klara's vision becomes "partitioned" when confronted with the Mother's perplexing human facial expression as she struggles to come to terms with the potentially fatal illness of her 14-year-old daughter Josie. The aesthetics of the novel is undeniably cubist, an unexpected departure from his earlier novels, where a surrealist's aesthetics dominates. For instance, in this other passage, the imagery of a woman being object-classified similar to a coffee cup soon gives way to a form of artificial hypercubism with the disfiguration of bodies,

"Then the Coffee Cup Lady reached the RPO Building side, and she and the man were holding each other so tightly they were like one large person." (p24)

Terms and phrases such as "partitioning," "a series of interlocking grids," "the Sun's pattern," "Open Plan," "blurred black-and-white pattern," and "rooms within rooms," allude to a new architectural aesthetics strangely mediated by Ishiguro's perceptive written language and Klara's perceptual vision system. Might Ishiguro's depiction of the hypercubist machine aesthetics be made generative for architecture? Is today's diffusion model the source of such aesthetics of disfiguration? Might Le Corbusier have extended this Artificial Hypercubism in the same manner as he did with Picasso's Cubism if he had lived in our age of AI? That is, reiterating it as his own Purism (e.g., "Drawing of Still Life," 1920-1922) and reappropriating it in the roof garden at his Villa Savoye (1929) as the so-called "fifth façade."

While all the works featured represent different experimental approaches in appropriating diffusion models for 3D architectural investigations, reflections, and designs, they are all based on Stable Diffusion v1, instead of MidJourney or DALL-E 2. Stable Diffusion model is the only opensource model, thus granting the opportunity in extending its text-to-image 2D diffusion model codebase with neural radiance fields as text-to-3D diffusion models for architecture. Unlike the pioneering work "3D-GAN Housing" exhibited at the Venice Architecture Biennale 2021 that demonstrated the use of deep generative adversarial networks (GANs) for generating semantic 3D models directly in high-resolution, these text-to-3D diffusion models do not require thousands of 3D CAD digital models as a training dataset, but rely on existing large and powerful pretrained diffusion models to hallucinate multiple 2D images from a given text prompt, and then optimize to reconstruct a 3D geometry based on their 2D view consistency. As a result, forms that emerge from such text-to-3D mechanism naturally lends themselves to a hypercubist aesthetics, or what is commonly known among generative AI researchers as the "Janus problem." Likewise, current text-to-image diffusion models are still struggling to generate coherent 2D imagery of hands and fingers, whose formal and spatial configurations are found to be apparently far more complex than those of faces. Such moments could be called "problems- turned- aesthetic" moments, which actually emerged in early GAN art imagery featuring displaced limbs and disfigured faces, as epitomized by the work of Mario Klingemann, such as his 2019 "Memories of Passersby I," then auctioned at Sotheby's. Artist Kyle McDonald in his 2018 article "How to recognise fake AI-generated images" is among the first to identify these GAN-generated artifacts of photorealistic faces, namely "missing earring," "asymmetry," "weird teeth," and "messy hair." Aaron Hertzman went further to argue that such visual indeterminacy is the very defining characteristic of GAN art, lying on the theoretical "uncanny ridge" afforded by "powerful-but-imperfect" GAN models. In retrospect, despite the recent shift from GANs to Diffusion models as SOTA, a recurring aesthetics of disfiguration and hypercubism remains somewhat visible, especially in 3D-Diffusion models.

©Immanuel Koh, The Hands of An Architect, 2023.

In illustrating traces of such disfiguration, the projects in the following pages will serve as 3D thought experiments and form studies: (1) 3D-Diffusion Pandas as Figure-Ground Body Studies, (2) 3D-Diffusion House as Volumetric Studies, (3) 3D-Diffusion House as Surface Studies, and (4) 3D-Diffusion Chair as Physical Studies. Searching for (or even verifying) the inherent neural aesthetics of 3D spatial and architectural forms.

©Immanuel Koh, 3D Janus Panda, 2023.

3D-Diffusion Pandas as Figure-Ground Body Studies

3D Prompt Neural Sampling series

Wu Guanzhong, one of China's greatest modern painters, is known for his abstraction and fragmentation of forms with ink on paper. His Pandas (1992), now part of the collection at the National Gallery Singapore, features a group of giant pandas that seem to morph into one another, whose distinctive black- and- white coat further heightens the effects of bodily fusion and disfiguration when depicted with his flowing black ink on paper.

Today's diffusion models are known to fail in accurately generating fingers and hands in 2D, and to produce the Janus problem in 3D. This study explores such AI "problems" as a possible aesthetic trajectory. Inspired by the figure-ground volumetric configuration alluded by the bodies of pandas, a text prompt is used as an input to a custom 3D-Diffusion model in kick-starting the sequence of denoising process, from an initial spherical noise to a hypercubist architectural configuration.

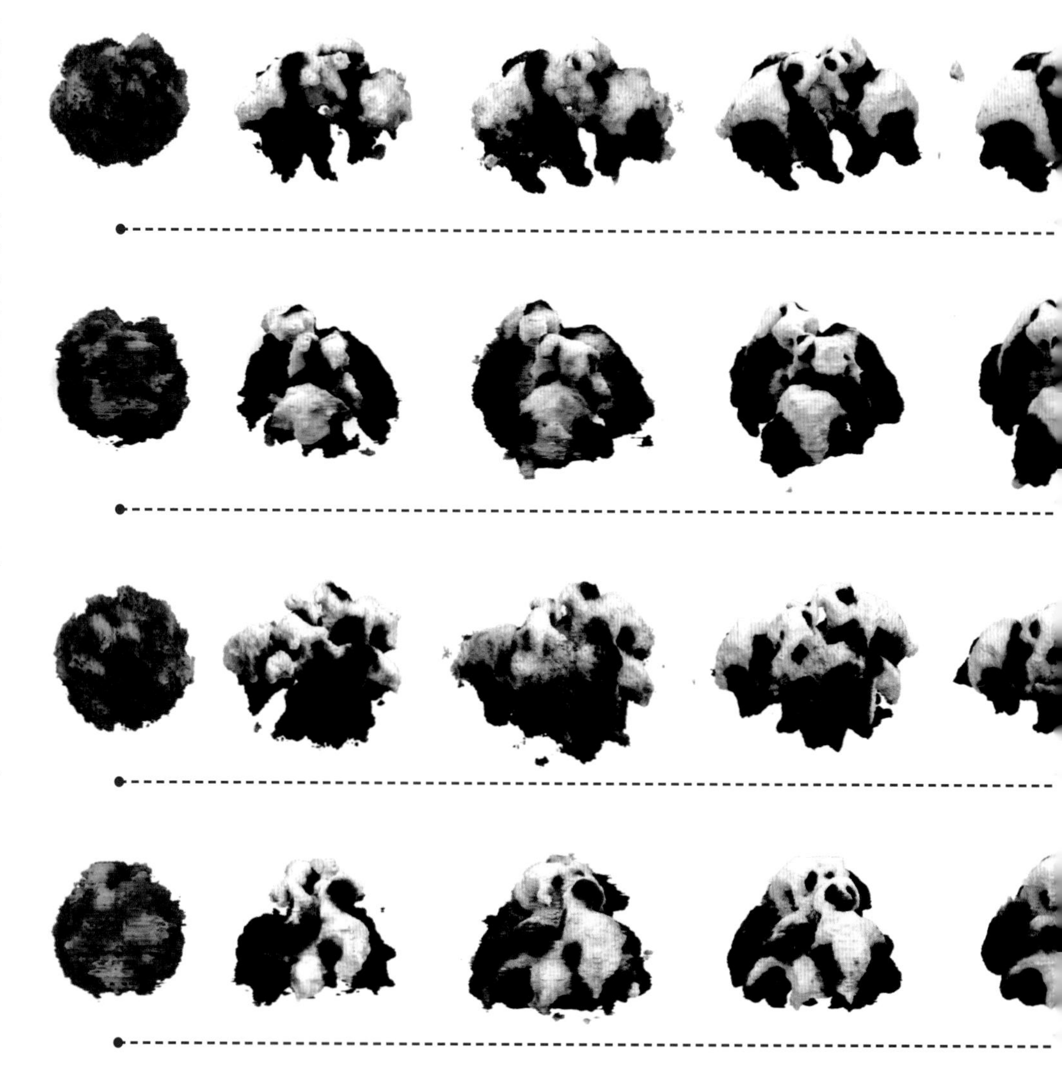

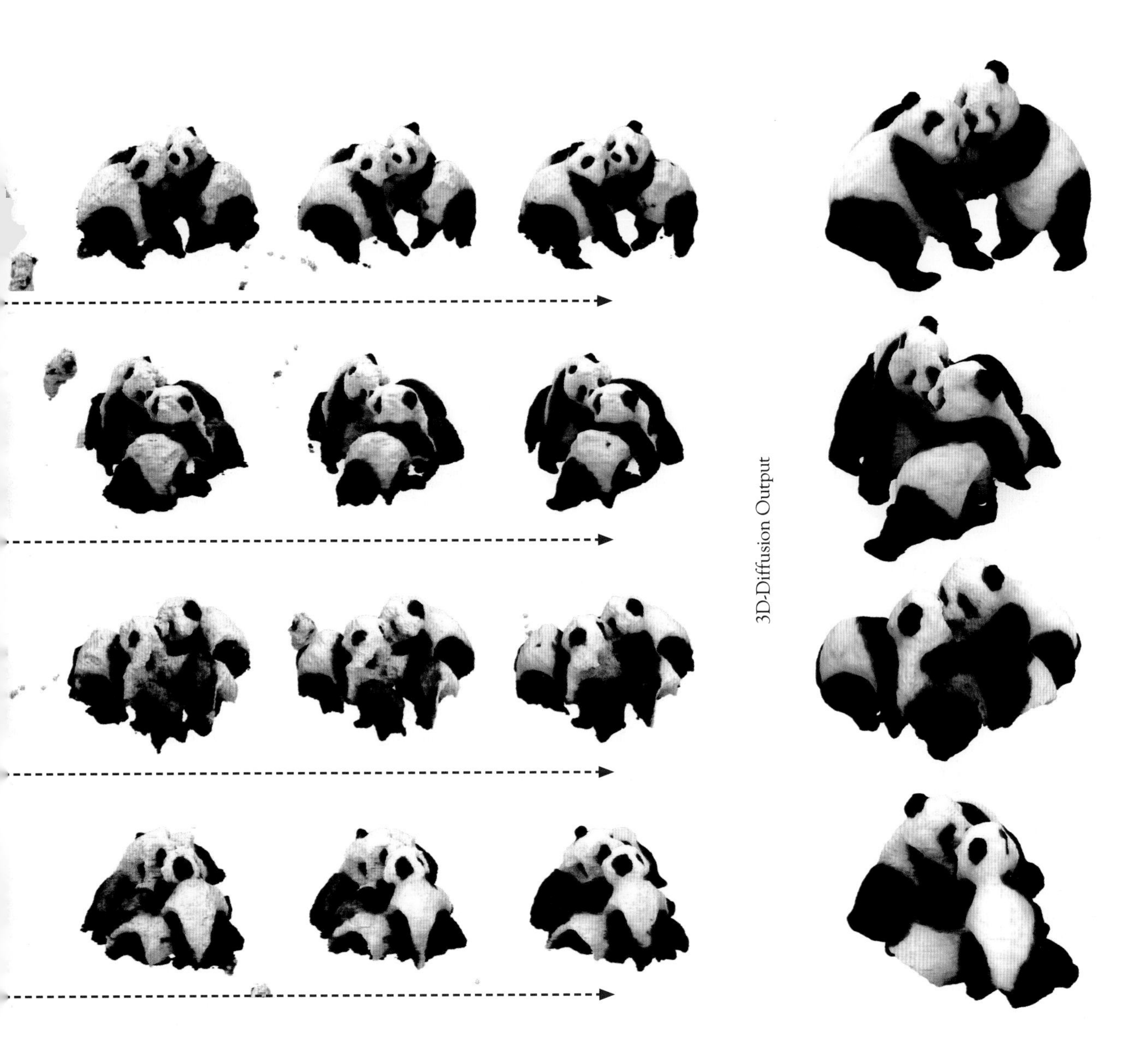
3D-Diffusion Output

3D-Diffusion House as Volume Studies

3D Prompt Neural Sampling series

Given a text prompt of "A house designed by _____," the 3D-Diffusion model synthesizes a 360° volumetric architectural form as well as its architectural site. The formal massing, spatial density, and textural colour information can be easily verified against the expected architectural styles of the respective architects Gehry's is characterized by the iridescence of the undulating titanium metal facade cladding and fragmentary volumetric interpenetration, recalling Guggenheim Museum Bilbao. Hadid's is characterized by the matte, smooth, curvilinear, and monolithic concrete volume and convex window apertures, recalling Heydar Aliyev Center. Kuma's is characterized by the aggregation of interlocking orthogonal discrete timber structures and traditional Japanese spatial articulation, recalling Yusuhara Bridge Museum. Yet, a hypercubist facade configuration seems to permeate all of them, resembling an image-like volumetric architecture.

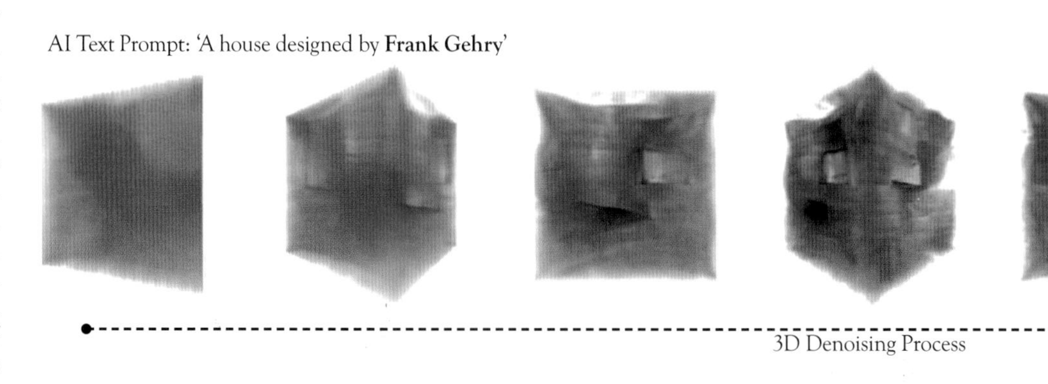

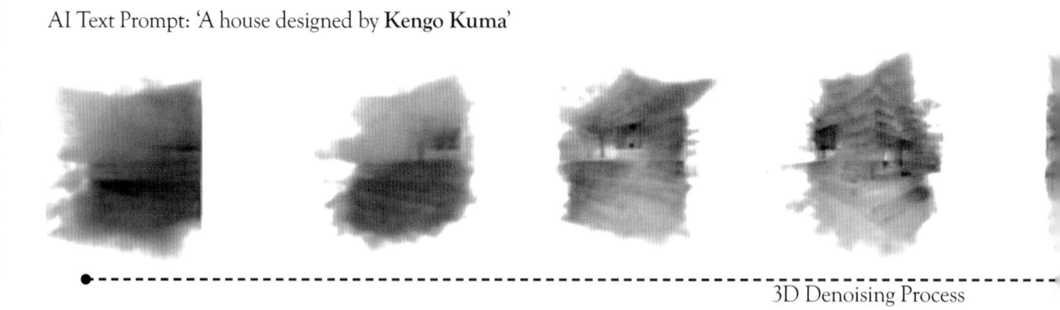

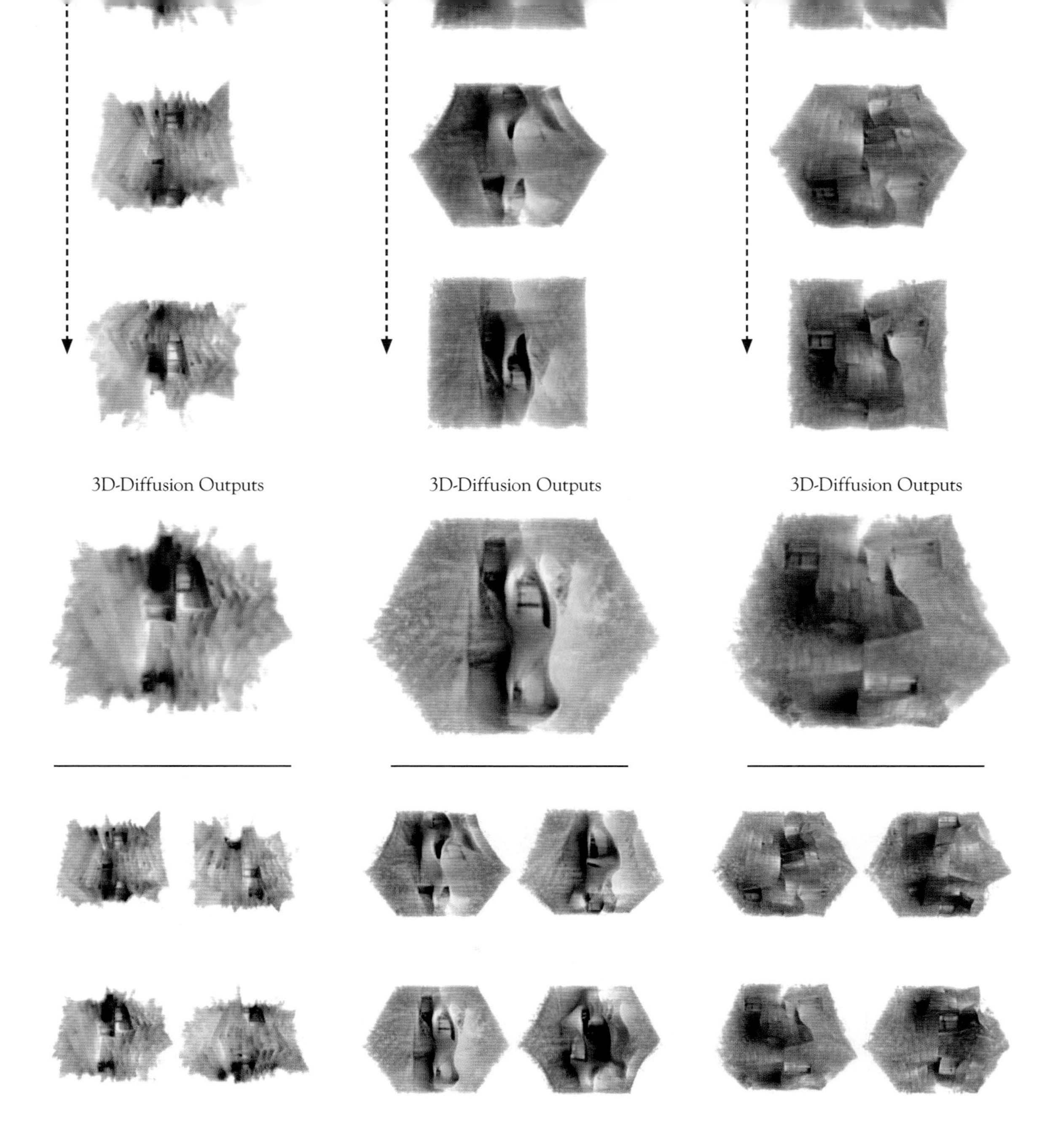
3D-Diffusion Outputs
3D-Diffusion Outputs
3D-Diffusion Outputs

3D-Diffusion House as Surface Studies

3D Prompt Neural Sampling series

Given a text prompt and an architectural 3D CAD model such as Le Corbusier's Villa Savoye, the model synthesizes all the surfaces of the architectural form directly in 3D, alongside its texture map. The generated textural colour information can be easily verified against the expected architectural styles of the respective architects. Gehry's is characterized by the iridescence of the titanium metal facade cladding and fragmentary volumetric placement seen on the roof garden, recalling Guggenheim Museum Bilbao. Hadid's is characterized by the matte monolithic concrete finish and curvilinear landscaping on the roof garden, recalling Heydar Aliyev Center. Kuma's is characterized by the discrete timber paneling on the facade and minimal intervention on the roof garden, recalling Yusuhara Bridge Museum. This study appropriates the generative potential of surface semantics which arguably first appeared in Picasso's Chouette Femme (1950) and Owl (1951).

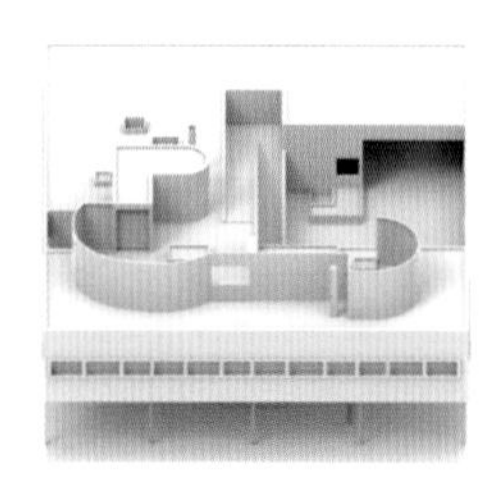
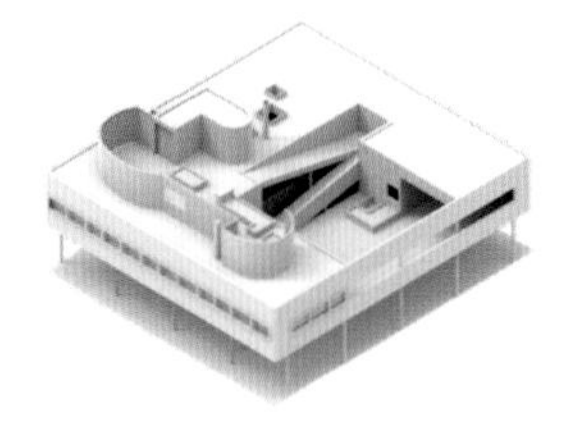
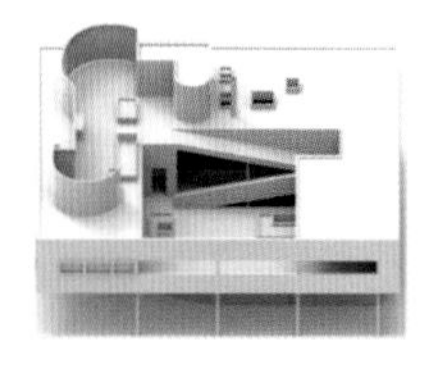
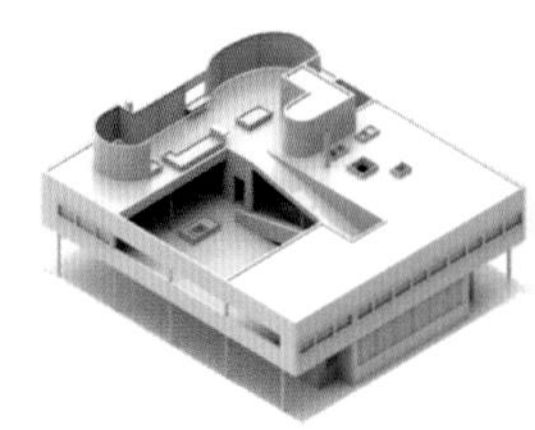

AI Text Prompt: 'A house designed by **Frank Gehry**'

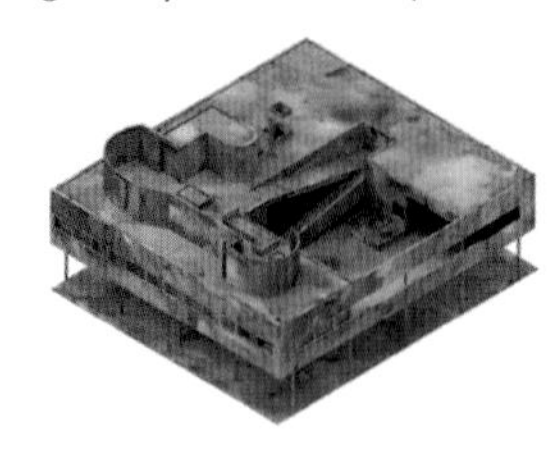

AI Text Prompt: 'A house designed by **Zaha Hadid**'

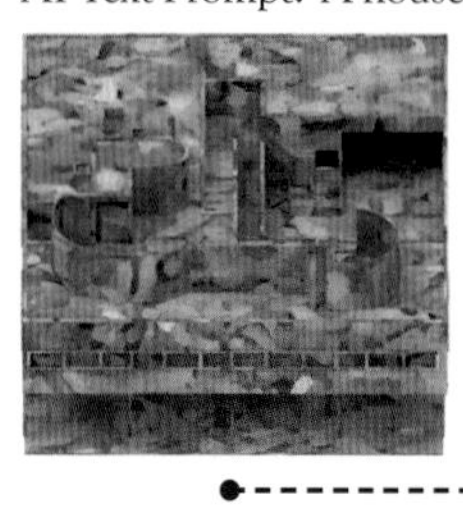
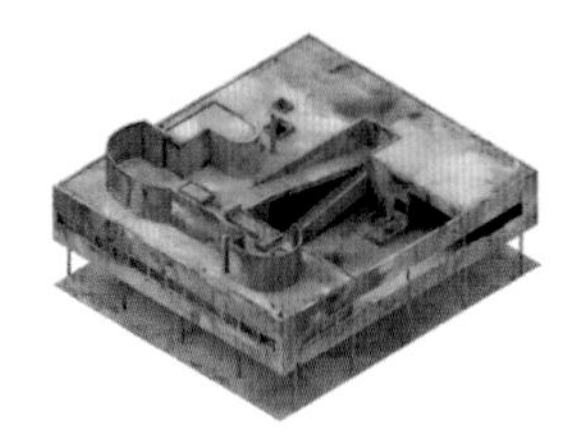

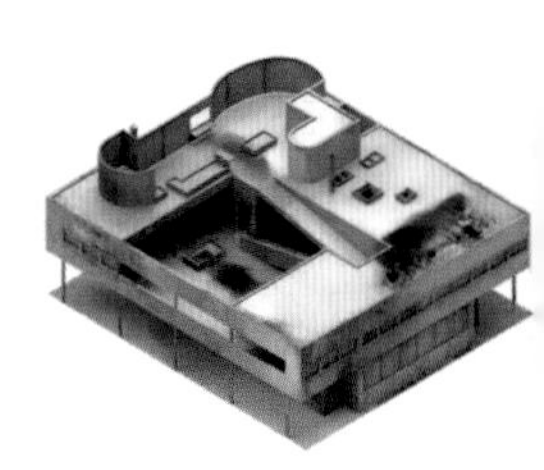

AI Text Prompt: 'A house designed by **Kengo Kuma**'

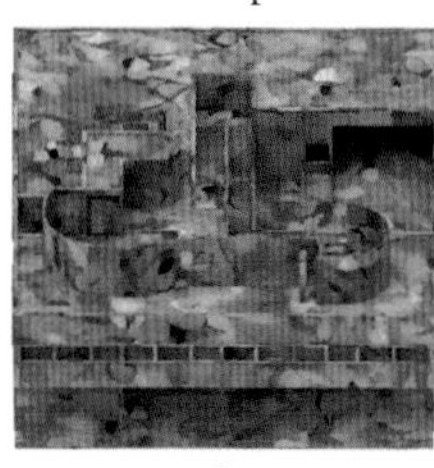
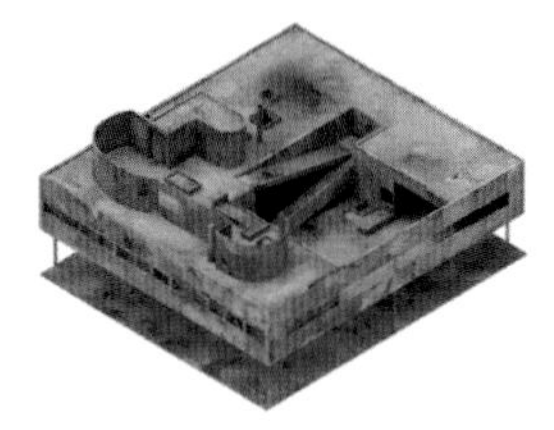

3D-Diffusion Outputs
3D-Diffusion Outputs
3D-Diffusion Outputs

3D-Diffusion Chair as Physical Studies

3D Prompt Neural Sampling series

Generative Adversarial Network (GAN) makes available a custom latent space for smooth interpolation in 3D after training it with a large dataset of 3D digital geometries as seen from the pioneering "3D-GAN-Ar-Chair-tecture" exhibited at the Venice Architecture Biennale 2021. The recent work "3D-GAN-Ar-Chair-teXture" (2022) is shown here as a latent walk sequence which is able to interpolate shape and texture latent spaces based on a training dataset of 3D-textured digital chairs. The second work shown here is a commission by Singapore's Asian Civilisation Museum in 2023 and forms part of the "3D-Diffusion-Ar-Chair-TEXTure" series. These generated white Janus-benches derive their styles from the museum collection of antique furniture by way of textual inversion. As a comparative study, the 3D-GAN-generated chairs exhibit a disfiguration aesthetics while the 3D-Diffusion-generated chairs reveal a stronger hypercubist aesthetics.

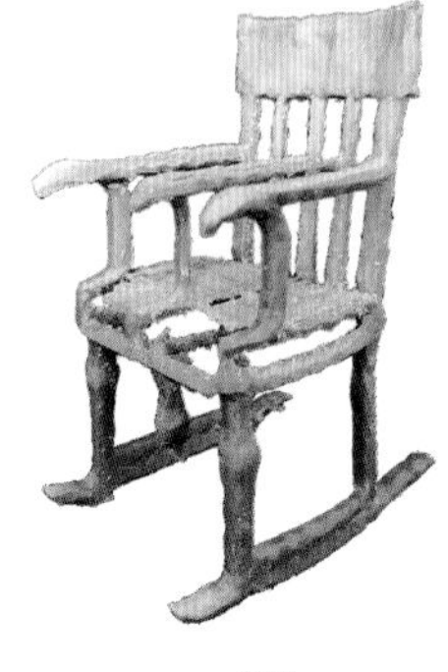

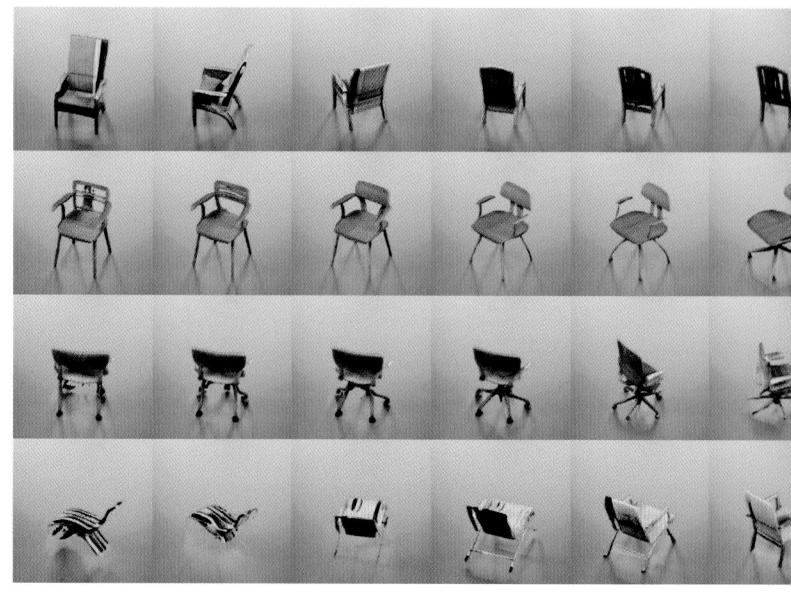

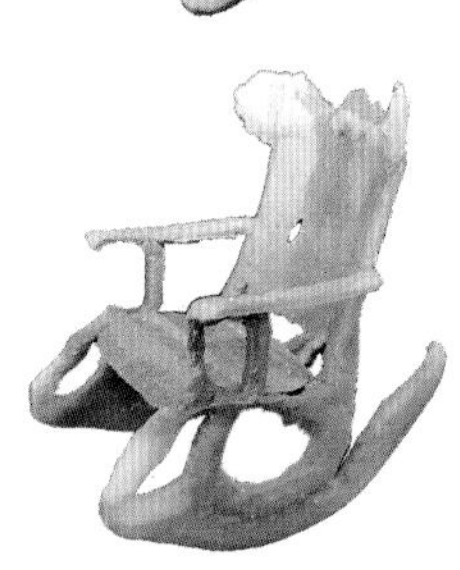

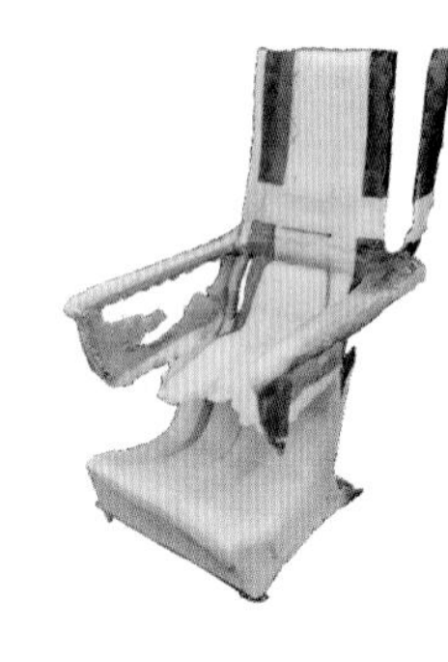

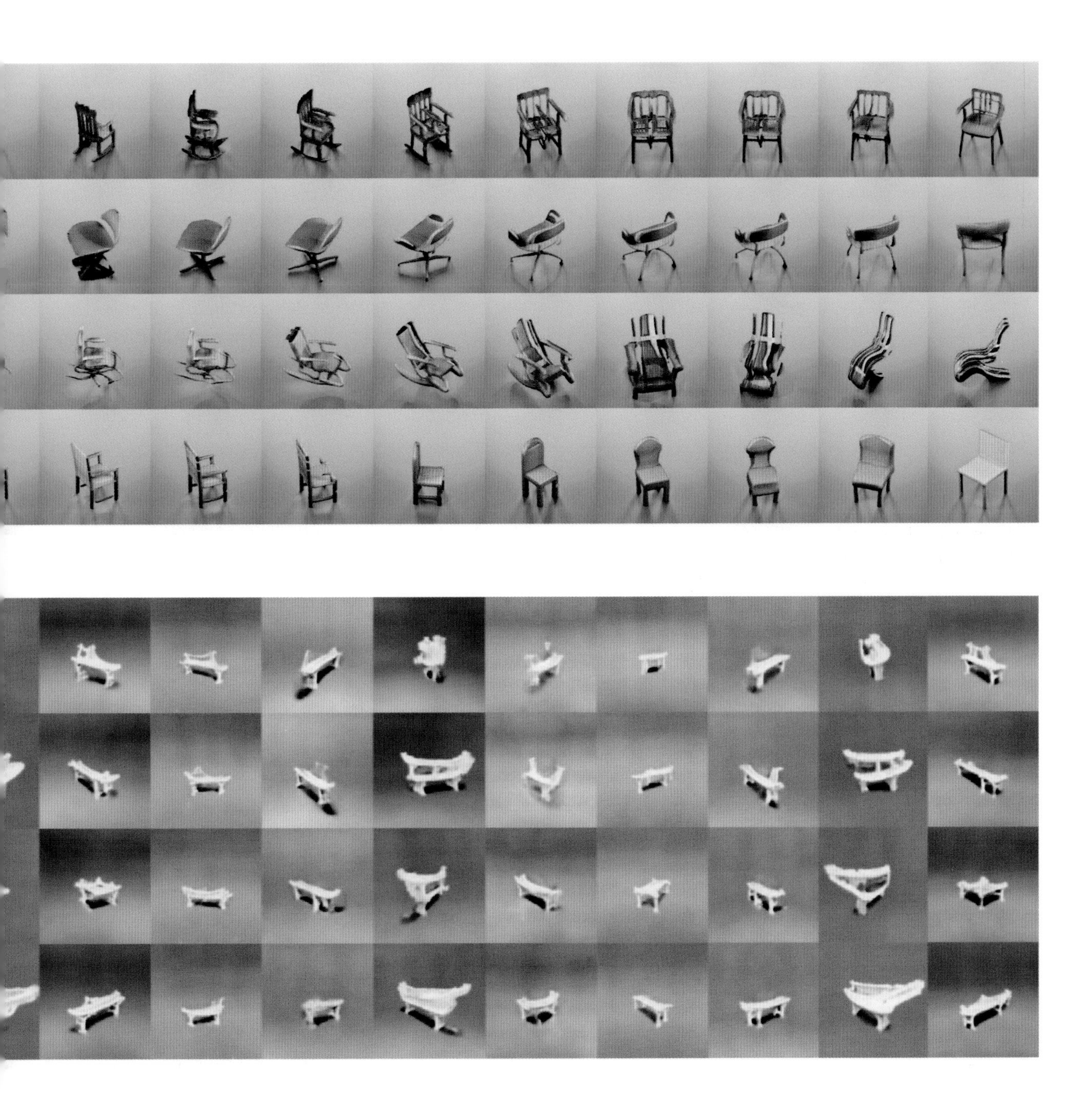

Role Play

Elena Manferdini

Identity is a mirror that reflects the different cultural and social factors that shape us as human beings. It refers to a person's sense of self, meaning how one views themselves as compared to other people. Identity has always been a shifting concept, not immune from fictional narratives, stereotypes, and appropriations. But the recent rise of our digital alternatives has drastically changed how we see ourselves.

Once upon a time, identity used to have a tangible manifestation because it could not be meaningful without the acknowledgment of others. Today, social media has fostered online communities. Our thoughts and most intimate feelings have been uploaded. Text messages, pictures, voices, and geographical coordinates have been compiled into an interconnected set of shared data that constitutes our digital neighborhood. Therefore, our sense of self is split between who we are in real life and how we represent ourselves online. Avatars, face filters, and deepfakes are just the beginning of the manifestations of a larger global transition of our sense of self into the cloud.

The digital turn has forced us to redefine our on-line persona, challenged the constructs of who we are as individuals, and granted all of us the possibility of being reimagined. Today, identity is filled with new risks and potentials that come along with role-playing. 2022 has been the year of text-to-image artificial intelligence (AI) tools. Midjourney, Disco Diffusion, and DALL-E have been spreading like wildfire, with a foreseeable impact on the current creative economy. Among these image generators, Midjourney differentiates from others because its visual results are "pretty by default" and capable of producing an overall fine art aura in a variety of styles. The tool is currently in open Beta and allows its users to feed in their prompts on Discord and then generate images akin to the text. Trained on a wide set of images of different art and architectural media, styles, and historic precedents, this tool associates reliably "aesthetic" images, often delightful and whimsical, to simple text prompts. Rapidly artists, architects, and designers have adopted Midjourney as an engine for imagination, along with a much larger community of users. "The tool is designed to unlock the creativity of ordinary people by giving them tools to make beautiful pictures just by describing them," explains David Holz, Midjourney Founder. And it is not so distant a future when visualizing one's imagination will be open-source, real-time, three-dimensional, and interactive. Design, as we know it will no longer belong to a relatively small group of well-trained individuals. Visualizing one's imagination will be a fully democratized practice, taking place in online communities where we are co-imagining. And these new human aesthetic open-sourced visions will spill out into the real world. And while this tool is without any doubt a game- changer for design, Midjourney visualizations have also proven to be a perfect mirror of an imperfect society. AI images are generated by models trained on a dataset of publicly available pictures amassed over time. Therefore, it is not surprising that they proliferate images that are dominant in our culture along with inventive new aesthetics. In short, AI image generators could be as racist or misogynistic as the data base they are trained on.

You and AI

The figurine is a hybrid of a robot from the future and rococo furniture from the past. The inanimate objects spring to life and assume anthropomorphic resemblance. The non-fungible tokens (NFTs) comes as a complement to the 3D-printed, hand-painted full-scale sculptures for use in the real world.

The project is an investigation on the role of artificial intelligence in contemporary imagination.

In 1993, Glen Ligon produced a series of lithographs titled The Runaways. Not dissimilarly from Midjourney - an AI program that creates images from textual descriptions- Glen's prints stemmed from word prompts; the artist asked a group of friends to write descriptions of him as if they were reporting a missing person to the police.

RAN AWAY, Glenn, a black male, 5'8", very short haircut, nearly completely shaved, stocky build, 155–165 lbs., medium complexion (not "light skinned," not "dark skinned," slightly orange). Wearing faded blue jeans, short sleeve button-down 1950s style shirt, nice glasses (small, oval shaped), no socks. Very articulate, seemingly well educated, does not look at you straight in the eye when talking to you. He's socially very adept, yet, paradoxically, he's somewhat of a loner.

RAN AWAY, a man named Glenn, five feet eight inches high, medium-brown skin, black-framed semi-cat-eyed glasses, close-cropped hair. Grey shirt, watch on left hand. Black shorts, black socks and black shoes. Distinguished-looking. The descriptions Glenn received read like accounts of criminal suspects, or racial profiling by law enforcement. The artist drew parallels between the texts his friends wrote, and the 19th- century advertisements adds published by slave owners to locate runaway slaves. Ligon explained, "'Runaways is broadly about how an individual's identity is inextricable from the way one is positioned in the culture, from the ways people see you, from historical and political contexts." The piece raised questions about racial identity and became a visual

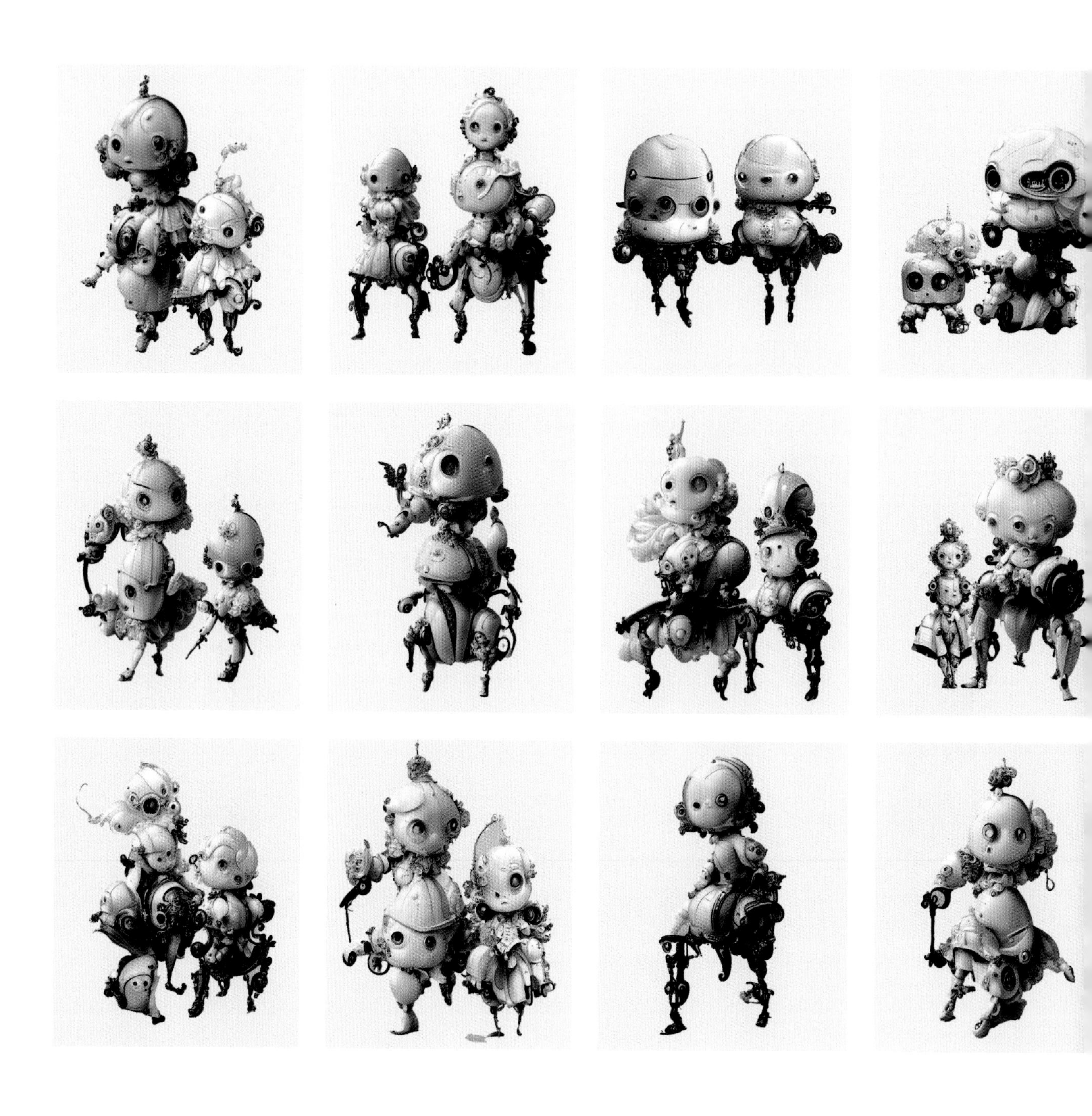

estrangements, role play, elena manferdini

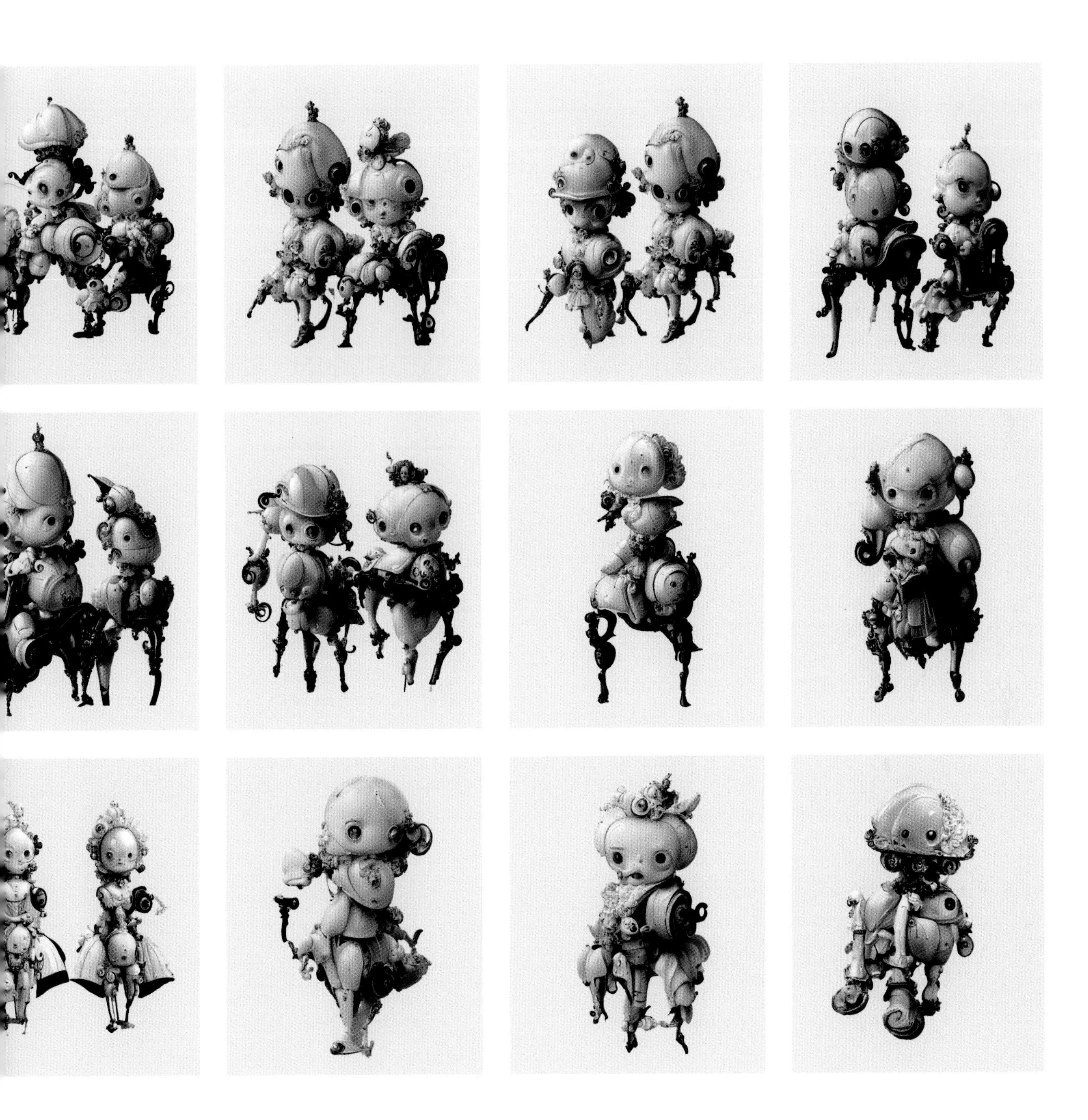

The series of 100 AI- generated NFT images allows collectors to own a unique, one-of-a-kind piece of digital art that can be bought and sold like traditional art.

manifestation of how far-reaching racial constructs and prejudice remain today.

The descriptions Glenn received read like accounts of criminal suspects, or racial profiling by law enforcement. The artist drew parallels between the texts his friends wrote, and the 19th-century advertisements adds published by slave owners to locate runaway slaves. The piece raised questions about racial identity and became a visual manifestation of how far-reaching racial constructs and prejudice remain today.

Another artist working with identity and dominant cultural stereotypes is Tomoko Sawada. The Japanese photographer shot myriads of self-portraits to reflect on women's identity and racial characterization. Using costume detail, hair, and makeup, Sawada constructed a prolific photographic imagery of herself. Her matrices of faces from afar appear as walls of conformity. In front of her work, spectators are uncertain if these pictures belong to one or more women, bringing to the fore the risk of racial generalizations. Her catalogues raise questions regarding women's individual expression and self-image in a society that educates them to conformity.

Glen Ligon and Tomoko Sawada make a case for cultural sensitivity and for a wider representation. Similarly to this body of work, Atelier Manferdini's Pick Me (2023) grapples with issues of identity at

The piece is a display of the ways AI tools are subtly proliferating dominant cultures within powerfully beautiful images. In particular the series of AI- generated portraits reflects on the effects of contemporary aesthetic judgments on female self-worth, and how media stereotypes connect aesthetic pleasure of cuteness with violence.

"Cuteness is a way of aestheticizing powerlessness. It hinges on a sentimental attitude toward the diminutive and/or weak, which is why cute objects—formally simple or noncomplex, and deeply associated with the infantile, the feminine, and the unthreatening—get even cuter when perceived as injured or disabled. So there's a sadistic side to this tender emotion," explains Sianne Ngai's in her book Our Aesthetic Categories: Zany, Cute, Interesting.

the time of an emerging AI The images could be loosely associated with a series of self-portraits. But they are not produced using a collage or a cell phone camera; they are generated by Midjourney departing from a single text input. The series is a temporal progression of faces. The first AI images are rooted in a cute Japanese cartoon style per text input. The last ones look like headshots of beauty pageant contestants. Scrolling through the faces, viewers notice that the age of the women oscillates from childhood to womanhood, and their initial manga cuteness transforms in to disenchantment, sadism, and substance abuse. One selection after the other, this AI- generated series has revealed how cuteness is an aesthetic commodity of the powerless in our culture.

Pick Me (2023) reflects on the effects of contemporary aesthetic judgments on female self-worth, and how media stereotypes connect aesthetic pleasure with violence. The piece is a display of the ways AI tools are subtly proliferating dominant cultures within powerfully beautiful images. "The man-made unfiltered data sourced online introduced the model to the existing biases of humankind. Essentially, A.I. is holding a mirror to our society." Explains Prisma Labs about the existence of bias in Stable Diffusion. While it's impossible to fully eliminate human bias from human-made tools, it's important to question if AI- generated images might end up reinforcing a dominant culture while they simultaneously produce speculative visualizations.

The Doghouse: Exhibition Installation, MAK Vienna, Austria

Sandra Manninger

The Doghouse is an installation conceived for the exhibition/Imagine - A Journey into The New Virtual at the MAK in Vienna, Austria. The exhibition presents new positions that deal in different ways with architecture and urban planning in the context of novel and advanced technologies such as Augmented Reality Virtual Reality (ARVR) and artificial intelligence (AI). Utopian, critical, futuristic and playful design strategies form the basis for new narratives, perspectives, and possibilities for action in virtual space that can continue into physical reality.[1]

At the heart of the installation is the idea that meaning is constructed through the use of signifiers- symbols, codes, protocols, and other forms of communication that we use to convey meaning. The use of three-dimensional (3D) modeling, robotics, and image generation all involve the use of signifiers, each contributing to the creation of a larger narrative about the intersection of sensibility and technology. However, the use of AI in the creation of this installation presents a particularly intriguing problem. As AI systems become increasingly sophisticated, they are capable of analyzing and interpreting data in ways that were previously impossible, leading to new forms of creative expression that challenge traditional notions of artistic authorship and intent.[2-3-4]

"The Doghouse" is an installation that speaks to the interplay of various signifiers, drawing attention to the complex web of meanings in the context of AI-generated content and human perception. The installation is composed of two distinct parts, a large-scale model of a house and a corresponding print, that work together to create a unified visual experience. The 3D model of the house is based on sections generated in the diffusion model Midjourney. Through the use of pixel projection, the 2D images generated by the Midjourney model are transformed into a 3D representation as a mesh model that is both precise and intricate. The resulting structure turns into a signifier in and of itself, a visual representation of the mathematical processes that underpin its creation. In addition, it addresses one of architecture's mainstays: designing 3D space through sections. In doing so, the installation serves as a commentary on the current condition of the Albertian Paradigm and its significance (or insignificance) for contemporary architecture that engages with machines capable of learning. For this, Alberti's idea of lineamenti[5] might be useful. Leon Battista Alberti introduced the concept of lineamenti in his treatise Della Pittura[6] (On Painting) in 1435. According to Alberti, lineamenti refers to the underlying geometric structure that supports the visible appearance of an object or form. When speaking about AI, the idea of lineament can be applied to the use of algorithms and code to create the underlying structure of digital images and objects. Just as Alberti believed that the lineament provides a foundation for the visible appearance of an object, AI algorithms can provide the underlying structure and logic that support the creation of digital images and objects.

To create the 3D model of the house, pixel projection was used. This method turns 2D images generated by the Midjourney into a 3D model. Pixel

The exhibition hall of the MAK located at the Weißkirchnerstraße.

The Doghouse - Midjourney Sections

Section studies for *The Doghouse* (SPAN 2023)

The Doghouse - SPAN (Matias del Campo, Sandra Manninger) 2023. Iteratively working through the problem of the section.

Sony Aibo (Artificial Intelligence Bot) Fourth Generation model ERS-1000

projection is a technique used to create 3D models from 2D images. It involves projecting each pixel in an image onto a 3D voxel grid to determine the presence or absence of a voxel at that location. To create a 3D model using pixel projection, the first step is to acquire a set of elevations, sections, or plans in 2D. Next, the images are processed to extract relevant features such as edges, corners, and textures, which are used to identify corresponding points across images. Once corresponding points have been identified, pixel projection is used to assign each pixel in an image to a 3D voxel in a voxel grid. This is achieved by computing the ray that passes through the camera focal point and the pixel coordinates in the image plane, and then intersecting the ray with the voxel grid. If the ray intersects a voxel, that voxel is marked as present in the 3D model, otherwise, it is marked as absent. After all the sections and plans have been processed in this way, the voxel grid is populated with voxels that represent the 3D shape of the object. The resolution of the voxel grid determines the level of detail of the resulting 3D model, with higher resolutions producing more detailed models. The resulting model was chopped up into pieces that could be Computer Numerical Control (CNC) milled, creating a precise and intricate structure. The Sony AIBO[7] robots that inhabit the 3D model of the house are another layer of signification, adding a playful and interactive element to the installation. The robots move and interact with each other and their environment, creating a dynamic and ever-changing scene that is constantly shifting in meaning. The robot's perception - it's machine vision- is utilized to transfer the perspective of the robot to a large audience via a website that transmits what the robots are seeing. This allows us to expand the effect of the installation beyond the boundaries given by the museum's space.

The large-scale print that accompanies the 3D model is equally significant, calling attention to the way in which architecture is composed of many discrete elements that come together to form a unified whole.[8] It discusses the part- to- whole nature of AI applications – from large- scale datasets of individual images (or any datapoints for that matter) to the restructuring of the data in the form of images, models, or text. The print appears to be a color gradient from a distance, but upon closer inspection, it is revealed to be made up of thousands of images generated by different image generators such as StyleGAN2,[9] Dall-E2,[10] Midjourney, and Stable Diffusion.[11] This complex mosaic of images becomes a signifier for the larger artwork, drawing attention to the intricate and multifaceted nature of the posthuman design ecology we are starting to experience.

The Doghouse {}

The Doghouse: Assembly sequence for the setup of the model in the Museum of Applied Arts (MAK) in Vienna, Austria. Drawing by Sang Wong Kang 2023.

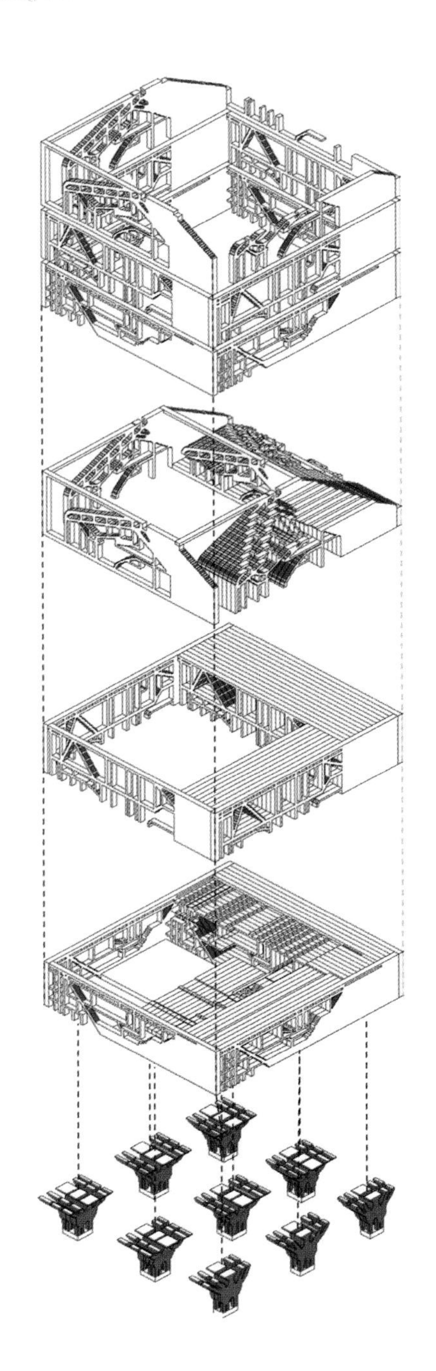

"The Doghouse" is an installation that speaks to the complex interplay of signifiers that surround contemporary architecture. Through its use of 3D modeling, robotics, and image generation, the installation becomes a visual representation of the various processes and technologies that underpin contemporary architecture production based on learning systems. As such, it invites the viewer to consider the ways in which architecture is composed of many discrete elements that come together to form a unified whole.

Roland Barthes believed that meaning was constructed through the interplay of signifiers, or the symbols and codes that we use to communicate.[12] "The Doghouse" is an installation

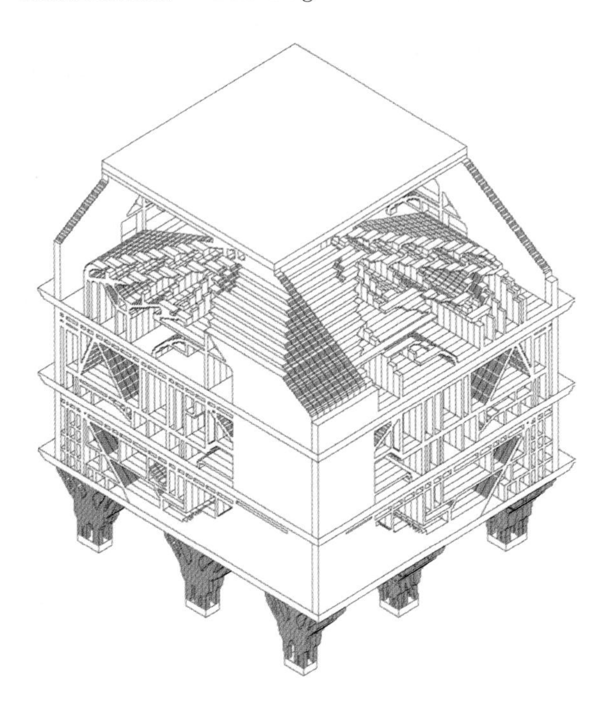

The Doghouse: Axonometric view of the assembled model. The model consists of 110 CNC milled components. Drawing by Sang Wong Kang.

The Doghouse: Fabrication of CNC - cut components for assembly in the MAK (File created by Sang Wong Kang) SPAN 2023.

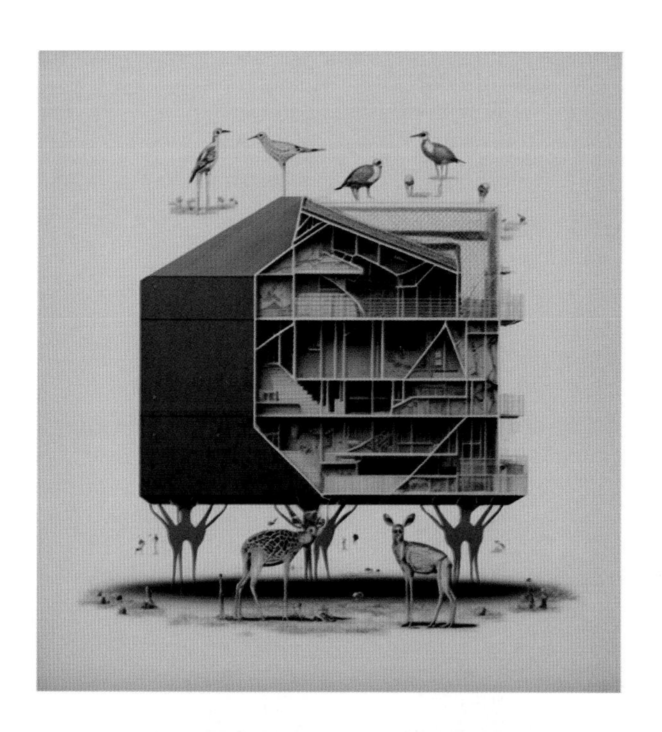

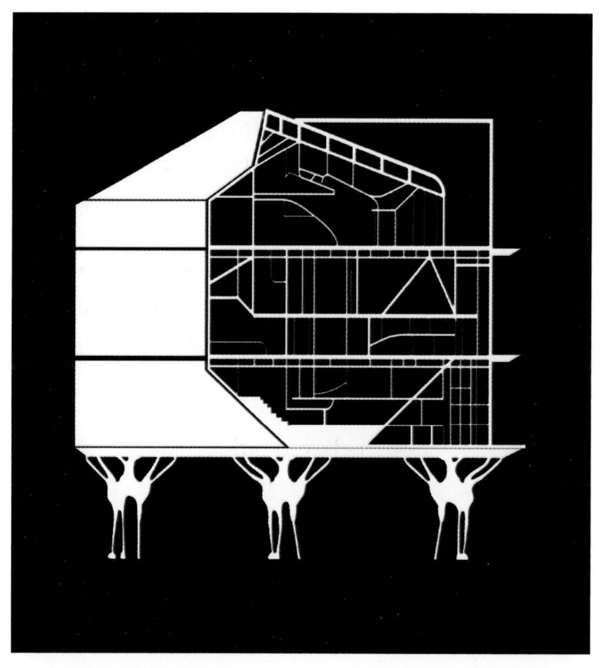

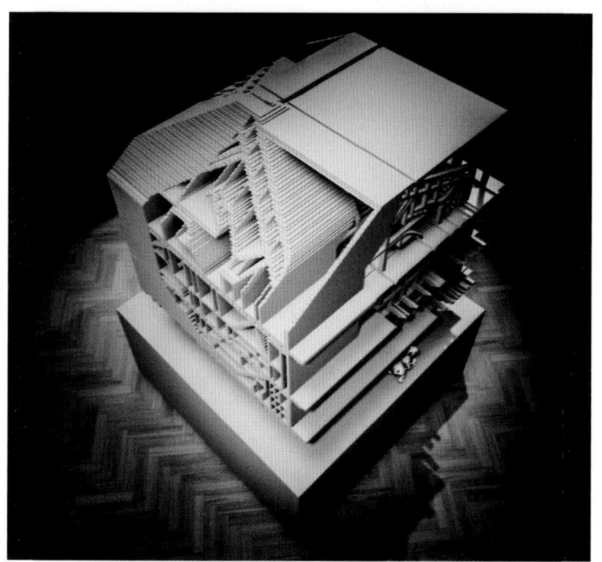

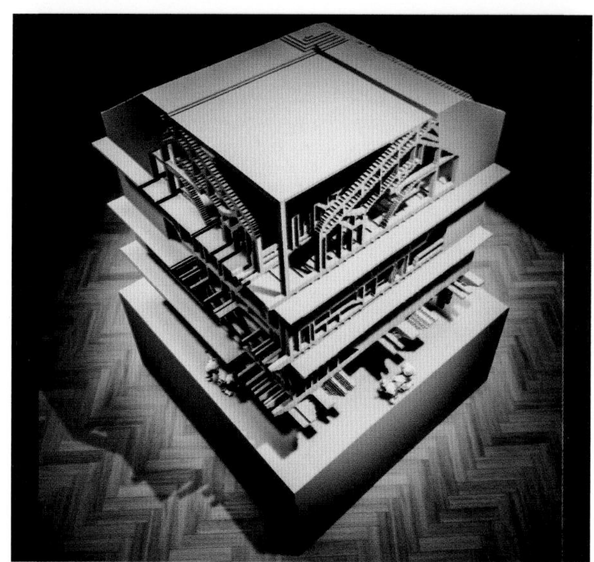

C

"Overall, 'The Doghouse' is an installation that speaks to the complex interplay of signifiers in contemporary architecture."

that embodies this idea, using various signifiers to create a complex and multifaceted work. The model of the house, for example, is a signifier for the underlying mathematical processes that went into its creation. Using pixel projection and CNC milling, the Midjourney model is transformed into a physical structure that is both precise and intricate. The resulting design becomes a visual representation of the complex and often unseen processes that underpin modern technology and artistic expression. In this context, the sign is the concept or idea being communicated, the signifier (the model) is the physical form in which it is communicated, and the signal is the actual transmission or communication of the signifier. All three of them are present in the installation – with the communication to the observer of the object being the most elusive one. At least in the physical space, in the virtual, it might be easier to measure.

The AIBO robots themselves are signifiers for the intersection of technology and humanity, creating a dynamic and interactive element to the artwork. As the robots move and interact with each other and their environment, they become a signifier for the ways in which technology can transform our understanding of the world around us. The gradient on the wall that accompanies the model is yet another layer of signification. From a distance, the print appears to be a color gradient, but upon closer inspection, it is revealed to be a mosaic of thousands of images. The use of these generators becomes a signifier for the ever-changing nature of contemporary architecture and technology. The gradient of the print is based on a Diffusion image. The feedback between the global appearance in the form of a color gradient and the generated 20.264 images the mosaic is made of, can in itself be read as a feedback loop between the information present in the generated images, and the data present in the billions of images used to train the diffusion model. To this end, it is calling attention to the way in which architecture is made of many discrete elements joined to form a unified whole.

The installation invites the viewer to engage with the artwork in a playful and interactive way, while also encouraging them to consider the complex processes and concepts that went into its creation. Through its use of 3D modeling, robotics, and image generation, the installation becomes a visual representation of the many different codes and symbols that we use to communicate meaning. By inviting the viewer to engage with these elements in a playful and interactive way, the work encourages us to question our assumptions about the nature of architecture and the ways in which it is created and understood.

However, the use of AI in the creation of the installation raises critical questions about the ways in which meaning is constructed and communicated. As AI systems become increasingly sophisticated, they can analyze and interpret data in ways that were previously impossible, leading to new forms of creative expression that challenge traditional notions of artistic authorship and intent.

"When speaking about artificial intelligence, the idea of lineament can be applied to the use of algorithms and code to create the underlying structure of digital images and objects."

The Doghouse installation in its context: the exhibition hall of the MAK in Vienna

C

The prompt that yielded the sections for *The Doghouse* evolved through a series of experiments. Incrementally improving and expanding the prompt, until a result was generated that allowed for further exploration:

offset print of an architectural blueprint drawing, perpendicular cross section of a modern house with a flat roof, made of giant birds nests, straw and skinny insects, bold zebra pattern, delicate thin elements, complex raumplan.

It took approximately 50 generations of images until the criteria necessary for the process of transforming the sections into a 3D model were met.

Project team:
Sandra Manninger, Matias del Campo, Sang Wong Kang, Brendan Tsai, Aditya Jaiswal, Devishi Kambiranda, and Chandana Rao.

The Doghouse
SPAN 2023

Mosaic - a gradient made of 26.264 individual images
estrangements, the doghouse, sandra manninger

Do Humans Dream of Furry Houses?

Vírgínia San Fratello

Furry Futures

There is an anecdote about the time Salvador Dalí gave Le Corbusier a tour of Gaudí's buildings in Barcelona. Naturally, Le Corbusier disliked them and gave Dali an impromptu lecture on purism and his *five points of a new architecture.* Dalí listened patiently, then said, *No. Architecture should be soft and hairy.*

Chinchilla Villa

Hair and fur are soft and comforting. Hairy or furry skin is hard-wired for petting and stroking sensations, creating intense pleasure when touched by both the receiver and the giver.

These artificial intelligence (AI)- generated furry home images elicit feelings of intense domestic pleasure and security. They demand cuddling and inspire comfort, coziness, and contentment. Each image is generated using the AI bot in Midjourney. Most prompts include a description of the house, "furry black chinchilla villa" or "alpaca mansion," in a landscape such as "the arctic tundra" or "Swiss alps" and a lighting scenario, such as "light halation, golden hour, soft lighting, glim lights, or volumetric lighting," as a way of moving from the descriptive text to the demonstrative image.

C

These furry diffusion models for homes, on the other hand, also refer to a very traditional method of construction called thatch, which can be seen in the photo below of a Nordic eelgrass roof.

Light halation. Soft lighting. Volumetric lighting. The golden hour. Backlit lighting. Uplight.

$$q(x_t | x_{t-1}) = \mathcal{N}(x_t; \mu_t = \sqrt{1-\beta_t}\, x_{t-1}, \Sigma_t = \beta_t I)$$

Diffusion Formula

A Fuzzy Pink Chinchilla Villa at the Golden Hour

Hunting for hairy huts in Hammerfest that are close to hot springs.

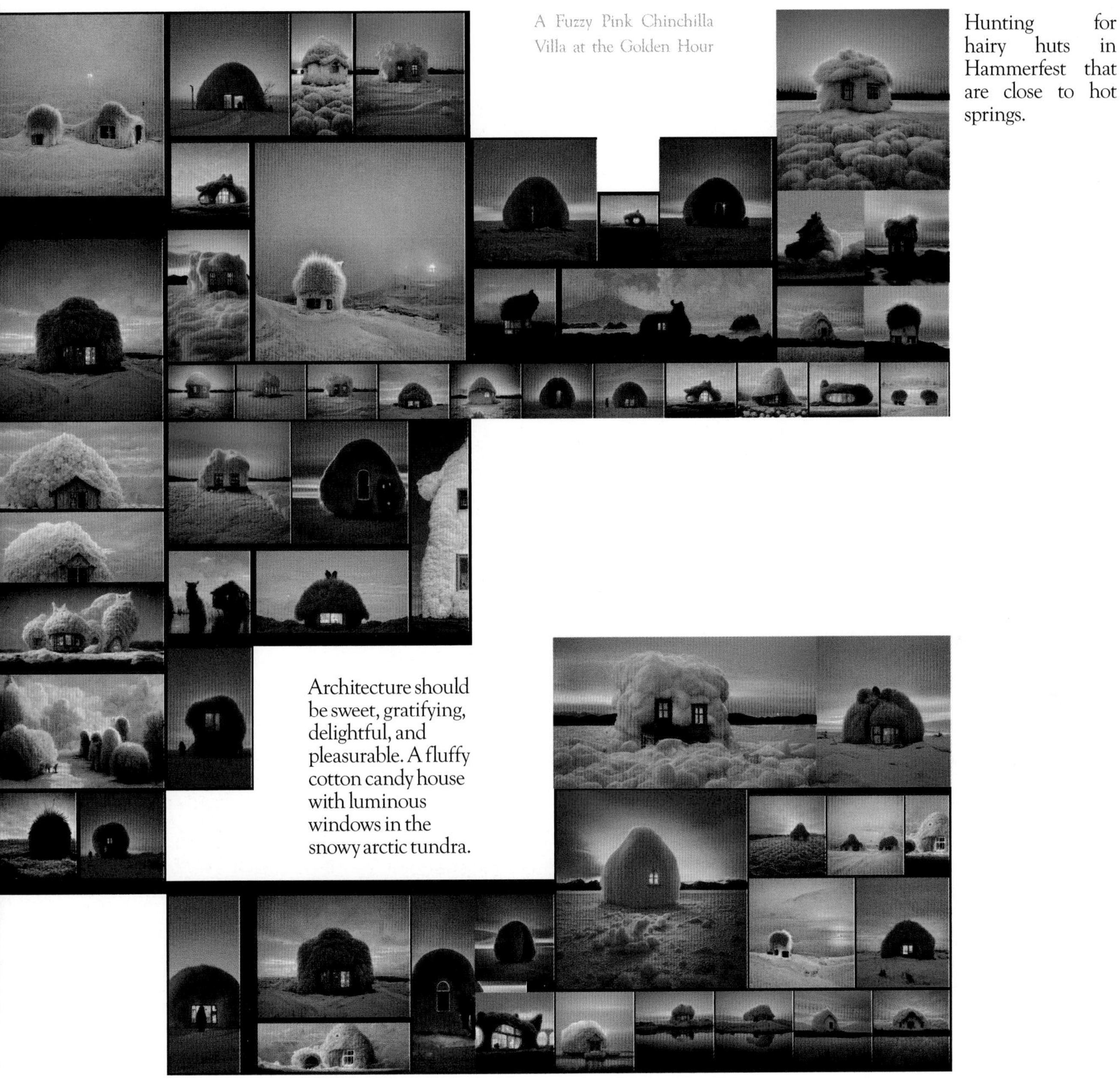

Architecture should be sweet, gratifying, delightful, and pleasurable. A fluffy cotton candy house with luminous windows in the snowy arctic tundra.

C

"Confession: I want to live in a furry house, one that I can cuddle, that will give me great pleasure and comfort."

{}

Animal Architecture

Zoomorphic architecture is the practice of using animal forms as the inspirational basis and blueprint for architectural designs. Biomemetic architecture uses processes from nature to help solve human problems. In these early AI sketches, furry and feathery animal species are combined with house typologies in order to speculate about new architectural forms, processes, and materials that address issues such as heat gain and loss, insulation, and sound attenuation in architecture. What happens when we combine a polar bear with a house? Polar bear hairs are hollow to maximize the insulating qualities of the animals' fur; if a building is also covered in hollow, clear tubes, could it be effectively heated by the sun like a polar bear is? Would homes that have long legs like a crane protect houses from high water or flooding? These sketches help us quickly imagine possible scenarios that could make buildings more adaptable to their climates and occupants. Most of the sketches generated as part of this series are from Midjourney versions 2 and 3, at a time when there was low coherence in the software and the resulting output was very creative and not very realistic, ideal for creating mashups between buildings and bears.

"Life creates conditions conducive to life." Janine Benyus

A mixture or fusion of different architectural and animalistic elements.

C

Bishon frise cabins, angora chalets, chinchilla villas, fox fur farms, alpaca abodes, and hairy houses all are parts of the texts that describe some of the animal architecture illustrated here. In some instances, it's not only the use of the animal name in the text prompt but also the reference to the author and illustrator Maruice Sendak that animalifies the buildings. The drawing style of Sendak makes these sketches wild imaginings filled with color, texture, and light and shadow. The sketches here draw inspiration from the hatching, line weights, patterns, and multiple light sources that can be found in Sendak's most famous book *Where the Wild Things Are.*

C

Pictures are often the simplest way to express the most complicated thoughts and feelings.

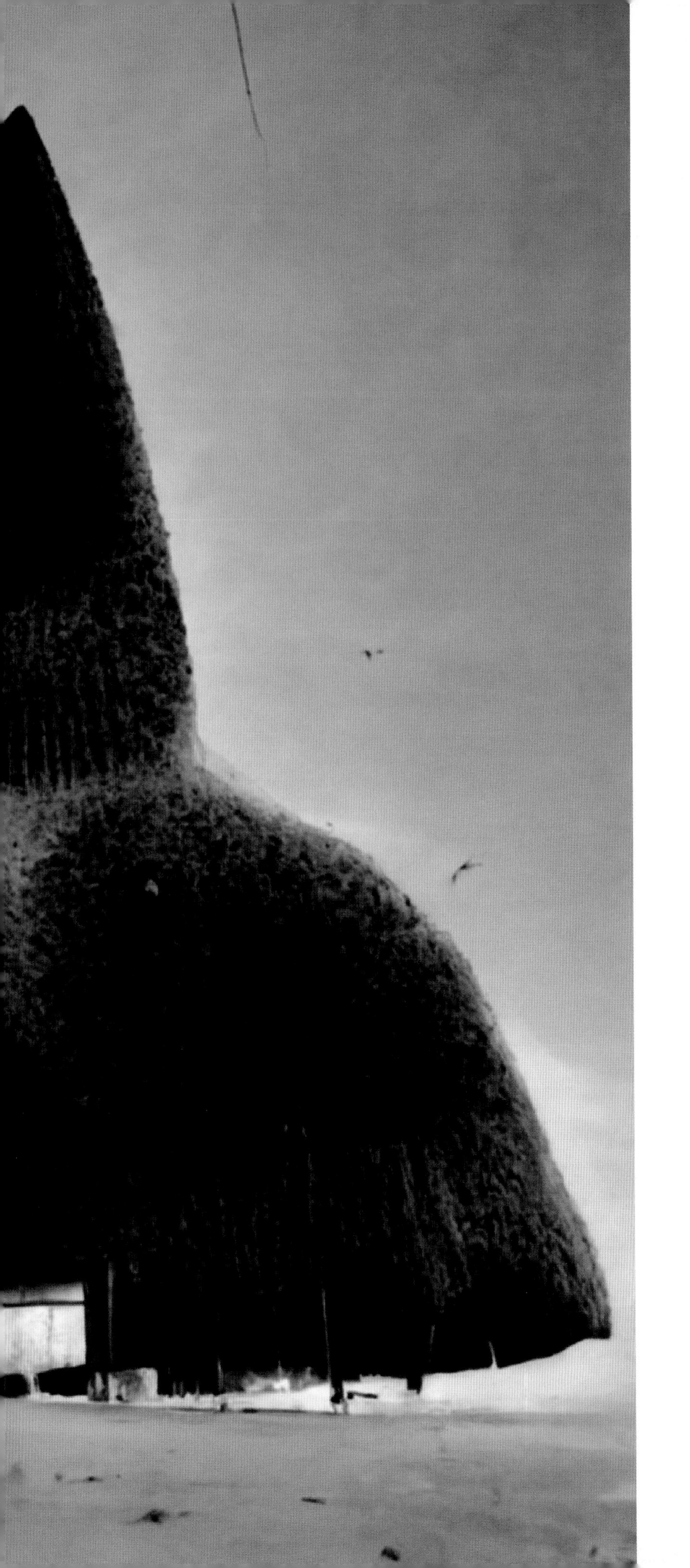

Allies in Exile

Kyle Steinfeld

A highly detailed professional photograph of a pilgrimage of Orthodox Jews praying outside a concrete wall, 3.0 blown snow in the Alaskan tundra, 3.0 Brutalism, 0.1 the Temple of Solomon, 3.0 Night sky aurora borealis, 3.0 asymmetrical composition, 5.0 huge crowds of Jews, 1.0 Monochrome, stark, awe, 1.0 worms eye shot, wide-angle, extreme panoramic, Dynamic Range, HDR, chromatic aberration, Orton effect, 8k, 1.0 Photo by Marc Adamus, Ryan Dyar, 0.5 --ar 16:9 --creative

Michael Chabon's 2007 novel *The Yiddish Policemen's Union* is set in an alternate history in which the US government provisioned land in Alaska in 1940 for the refugee settlement of European Jews fleeing Nazi persecution. With the fledgling State of Israel destroyed in 1948, European Jewry found a new home not in New York, but in Sitka, a large, Yiddish-speaking metropolis initially established as a small temporary settlement on unceded Tlingit tribal land. The plot centers on the illicit construction of an architecture, and unfolds within atmospheric world that blends Jewish and Tlingit cultures, in the Alaskan landscape. Thematically, this work, that is set in an alternative present explores a breadth of issues that resonate with our actual present, including: identity and community; extremism and assimilation; and cross-cultural contact, friction, and appropriation. Chabon's writing is vivid, and since reading this work, I've been a bit obsessed with his imaginary Sitka- something like a 19th - century Vienna in the Arctic. What would the material culture of this place be like? How would the trauma and traditions of people from the *old world* experience modernity in this unique climate and cultural situation? Would Jewish design and culture harmoniously meld with the indigenous traditions (Tlingit talit, anyone?), or would this alternate history rhyme with our own, with the transplanted heart of Jewish life beset by issues of exploitation, appropriation, and oppression?

I raise Chabon's opus of speculative fiction in this context for two reasons. First and Foremost, as you can likely tell, I'm a fan. This is evidenced by the nearby synthetic images - speculations set in

A highly detailed professional photograph of a pilgrimage of Orthodox Jews praying outside a concrete wall, 3.0 blown snow in the Alaskan tundra, 3.0 Brutalism, 0.1 the Temple of Solomon, 3.0 Night sky aurora borealis, 3.0 asymmetrical composition, 5.0 huge crowds of Jews, 1.0 Monochrome, stark, awe, 1.0 worms eye shot, wide-angle, extreme panoramic, Dynamic Range, HDR, chromatic aberration, Orton effect, 8k, 1.0 Photo by Marc Adamus, Ryan Dyar, 0.5 --ar 16:9 --creative

In the context of prompt-based architectural visualization, designers must learn to navigate landscapes beyond the technical.

Chabon's alternative present. These explorations *worldbuild* the architecture and artifacts of a place that, had history gone a slightly different direction, I might have been a citizen of. But beyond these largely personal indulgences in fan fiction, I would argue that Chabon's novel opens a useful conversation for all of us who operate at the intersection of generative artificial intelligence (AI) and architecture.

Working with image-synthesizing CLIP-guided diffusion probabilistic models, or *diffusion models*, feels different than CAD. It even feels different than other forms of architectural image-making. In contrast with traditional visualization, working in this way feels less like constructing visions of the future, and more akin to conjuring alternate versions of our present. Perhaps this is because the dazzling synthetic images that draw so much attention on Instagram are not de novo visions of the future drawn from the mind of a sole creative author. Rather, these are inferences drawn from data scraped from our collective past- data laden with the visual richness of, as well as the latent biases of, the cultures from which they were extracted. In this way, we are all engaging in a form of *fan fiction* at the moment we invoke in our prompts the proper name of an author, or an architectural style, or mention a specific geographical region or culture that manifests that unnameable quality that we seek. It's undeniable in this moment that we're working with a technology of culture, one that bears the indelible mark of the context that produced it. In stark contrast with traditional CAD, which assumes a posture of objectivity and neutrality,

A portrait of the back of a Jewish man in prayer from behind wearing a taillis with Tlingit ornament. The man is in the snow. Photograph by Annie Leibovitz. --ar 9:16

A professional photograph of Priests sacrificing a lamb in a very beautiful synagogue designed by a stealth fighter::4.0 Angular, minimal, stark, harsh, stealth, glossy white::3.0 Tlingit arabesque ornament::2.0 A crowd of Orthodox Jews::2.0 Very detailed, very ornamented, very arabesque::3.0 worms eye shot, wide-angle, extreme panoramic, Dynamic Range, HDR, chromatic aberration, Orton effect, 8k::1.0 Photo by Andres Gursky::3.0 --q 2 --ar 9:16

diffusion models insist that we address many of the same issues raised by Chabon's work: issues of identity, community, extremism, appropriation, and exploitation.

In my early engagement with these systems, I felt compelled to address the issues mentioned above, but also felt hopelessly underprepared to face them. With nothing in my education as a computational designer to guide me, I found myself instead turning to my personal experiences - to memories, stories, and cultural references that I knew well. The nearby images, for example, show scenes and forms drawn from my own childhood: nonce orders assembled from palmetto grass, temporary structures assembled from the wreckage of the Challenger disaster, and houseboats surrounded by manatees that can no longer survive in the wetlands of Northern Florida. In many ways, all of us in the design computation community are under-prepared to face the issues raised by this new digital *turn* - a term we use to describe moments of disciplinary crisis. The imagery that has dominated architectural social media across the past year was produced through methods that are unlike anything we've dealt with before. It seems we can forget everything we know about parametric data trees, object-oriented programming, and digital morphogenesis - **this is fan-fiction all the way down**. Such a shift holds narrow ramifications for those of us who work in design computation, and also broad implications for the discipline and the position of technology within it. Those of us in design computation have a daunting task in front of us. As stewards of the cultural and social dynamics of the discipline's

Photographs of a nonce order column capital made of alligators and dirt bikes and palm trees and glass bulb pipes and shit and shit, Sigma 75#mm, ornate, very detailed, very ornamented, intricate, studio lighting, wide-angle, polished, hyperrealistic, low angle, bokeh.

A NASA solid rocket booster floating in a swamp. An informal statement full of people –stylize 625 –quality 0.5–ar 16:9–ar 16:9

engagement with technology, it falls to us to provide a critical assessment of software tools employed by architects, a record of emerging methods of design enabled by these tools, and an accounting of how the affordances of these tools and methods intersect with creative practice. Our role takes on particular significance in periods of pronounced technological change, when the dynamics of tools and practice are seen in stark relief. In many respects, we've been here before. Many of us in this community

A photograph of a door knocker shaped like a bird designed by (from left to right, top to bottom) [Aldo Rossi, Filippo Brunelleschi, Renzo Piano, Carlo Scarpa].

A photograph of a modern home on a hilltop by the sea translated into (from top to bottom) [Korean, English, German, Spanish].

From top to bottom: images generated by the prompts: *Wood or Metal, Charlotte Perriand, 1929, Learning from Las Vegas, Denise Scott Brown and Robert Venturi, 1972, The Five Points of a New Architecture, Le Corbusier, 1927, With Infinite Slowness Arises Great Form, Ludwig Mies van der Rohe, 1953.*

served in such a capacity during the last digital *turn*, when parametric and generative design entered the architectural consciousness in the early 2000s.

This time is different, and we're not prepared.

The last time, our community called upon a specific constellation of allies to cope with disciplinary questions we couldn't adequately address on our own. In this earlier period, a specific set of organizations, events, and collectives were established (see, for example, the Smartgeometry group founded by Robert Aish and others in 2001, or the Institute of Computational Design at the University of Stuttgart) and served to bring architectural practitioners and academics together with a range of expertise not typically required

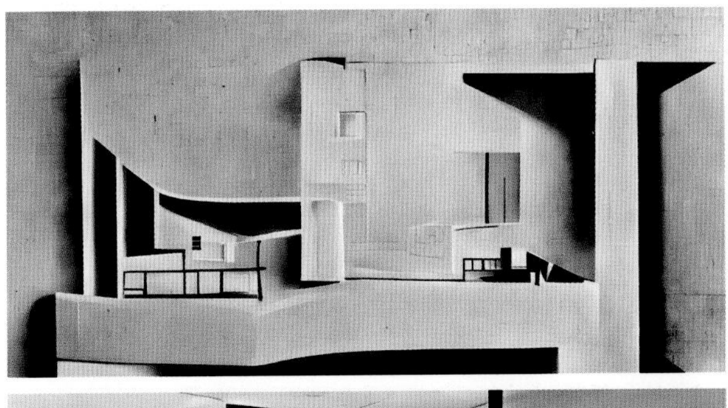

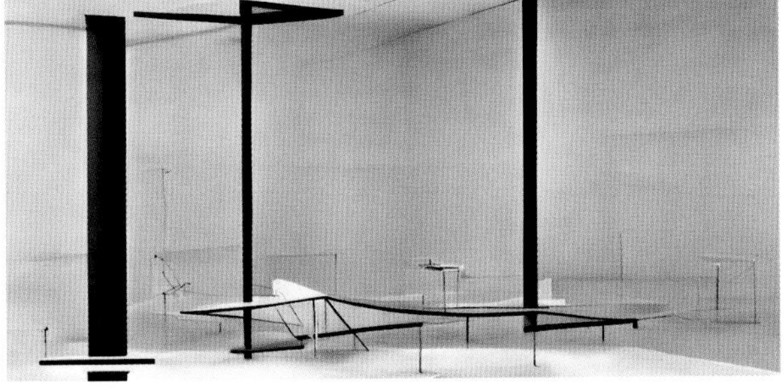

by our discipline. Given the technologies at play, this included some figures that we might expect, such as computer scientists, structural engineers, and professional programmers, as well as some we might not, such as systems theorists, computational geneticists, and artificial life enthusiasts. Such an eclectic constellation of voices was required in that moment in order to further the broader project, and address critical questions regarding the integration of a specific set of emerging and extrinsic design methods (e.g., optimization, emergence, and morphogenisis) into architectural practice.

This time - both in terms of the nature of the technology, and the nature of the surrounding questions - things are different. In some of the images shown nearby, I've sought to address some of the ways that issues of identity, appropriation, and culture are unavoidable when working with diffusion models. For example, since language itself is a cultural signifier, the series of images on the far left illustrates the unavoidable non-neutrality of text-to-image systems. Consider that the software that produced this set of images, each invoking the same prompt translated into four different languages, would operate differently for different users - it would perform differently in Kansas than it would in Korea or Columbia. Extending this small example to computer-augmented design more broadly, the implications are sobering. While our tools have never been truly neutral, how might design practice adapt to working with systems so deeply intertwined with culture, and that they can no longer be mistaken as neutral?

As in the previous turn, this new technology suggests a set of issues extrinsic to our discipline. Just as before, we will need allies, but the specific constellation will certainly be different. Some of these voices we require will be technical, including data scientists and machine learning engineers, but the most critical guidance may come from elsewhere, such as data ethicists, crypto artists, historians, and anthropologists. Two subject areas hold particular relevance, and are worth expanding upon: each emphasize the cultural role of computation and has historically considered *external to the discipline* of architecture. First, software studies is an interdisciplinary research field that accounts for the social and cultural effects of software systems and includes such figures as Benjamin Bratton and Lev Manovich (a contributor to this text). While a better connection with this community is long overdue in design, an approach that understands CAD through the lens of media - that is, as an instrument that connects us to knowledge, skills, and abilities that we would not otherwise hold - would prove particularly useful in this moment. Further, figures versed in digital labor studies, such as Trebor Scholz and Niloufar Salehi, would help us unpack the techno-social power dynamics at play in this new landscape, including issues

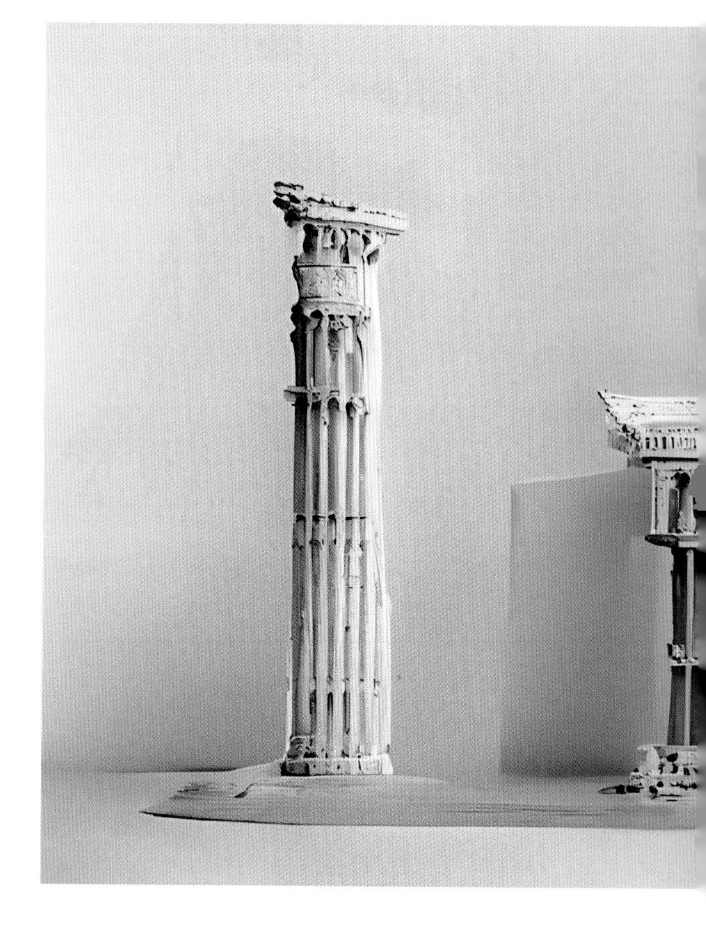

An image generated with the prompt: *De Architectura, Marcus Vitruvius Pollio, 20 BC.*

of exploitation and appropriation surrounding datasets. In many ways, the introduction of text-to-image into architectural visualization work represents a classic technological labor disruption of the sort that produces clear socioeconomic *winners* and *losers*. While typically the winners in such a scenario tend to be the tech-savvy, the dynamics here may be different in that the lowering of barriers has already flooded our market with compelling and imaginative images, each of which was made

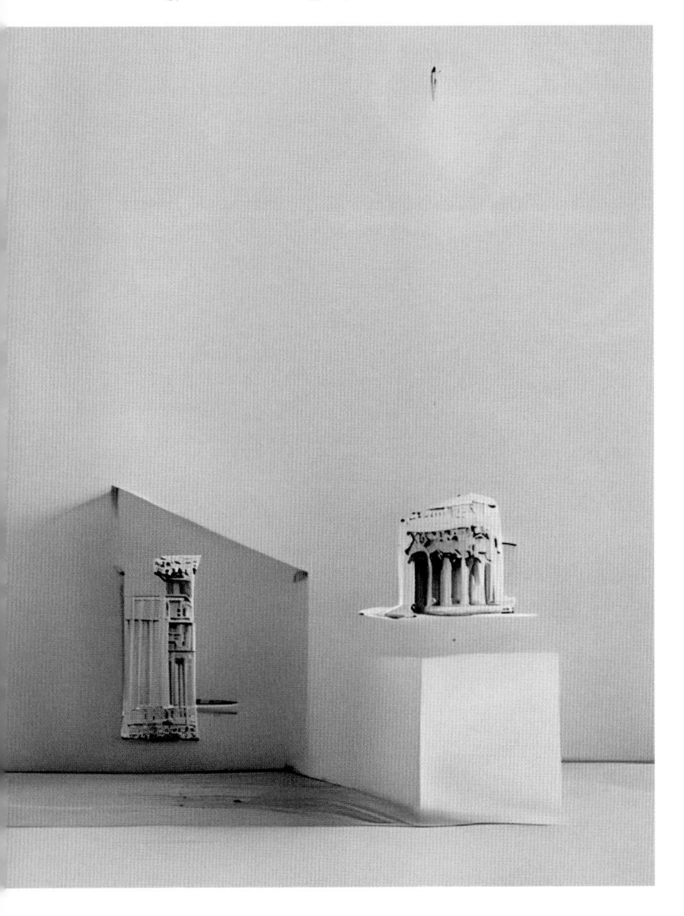

in seconds. As any labor economist will tell us, when a product becomes much easier to produce, this project is devalued. It would seem that we're entering a phase in which architectural image making is easy, cheap, quick, and potentially holding radically less value than it has held in the past. How will the flood of incredibly beautiful and radically cheap images change our discipline?

Like Chabon's refugees exiled in Sitka, there are those of us in design who may feel suddenly thrust into a foreign landscape, daunted by the technical complexities of a new technological domain, and unprepared for the ethical implications thereof. It falls to those of us who advocate for the humanistic applications of technology, together with the allies mentioned above and others, to assist practitioners in critically and ethically engaging with this powerful new computational *tool of the imagination*, and to help foster positive cultures of practice around it. A central question remains: to what extent will ML tools catalyze a shift in the centers of power in design practice? Along with clients, contractors, developers, regulatory bodies, architects are but one small part of a broader network of social, technical, financial, and cultural exchange. In a context in which architectural labor is already beset by inequities and exploitation, how might the seemingly inevitable changes on the horizon exacerbate these existing problems?

Or, worse yet, might things not change at all? Might we miss a rare opportunity for transformation and fail to bring the discipline more in line with our values?

///// APPENDIX /////

Epilogue

Matías del Campo

There are probably a million ways to end this book, but let me end with a little story. Walter Gropius could not sketch. At all. Not a bit! At the start of his career working in Peter Behrens' office, Gropius kept this handicap a secret.[1] Despite this setback, Gropius persevered in the Behrens office and eventually became known for his ability to dictate drawings to his collaborators, proving the power of language to convey the intricate material and spatial relationships in architectural projects.[2] This notion of description was further expanded upon by Sol LeWitt, who dedicated his entire career to the potency of language in the form of instructions, presaging the emergence of programming and scripting as a means of artistic and architectural expression.[3] The synergy between language and architecture is complex and multifaceted, yet it is clear that language has the ability to convey abstract concepts and facilitate collaboration, offering a powerful tool in the design process. To this end, this book attempted to shed light on the murky shadows of diffusion models by exploring their relationship with language and the architectural imagination. As we close the last chapter of this book on diffusion models in architectural design, it seems unmistakable to contemplate on the ways in which language, prompting, and the theoretical perceptions of thinkers like Wittgenstein, Roland Barthes, and Michel Foucault have fashioned our views upon technologies aiding in the imaginative process. At the very beginning stands Wittgenstein's quote "the limits of my language is the limits of my world," so, what do we know now? After plowing through a plethora of projects using tightly related methodologies of teasing imagery out of diffusion models? What is the epistemology of diffusion models in architecture? What would Wittgenstein say about large language models and diffusion algorithms?

Language, Authorship, and Estrangement

Of course, it is purely speculative, but making deductions from his writings one could argue that Wittgenstein would have firmly suggested that the meaning of language is closely tied to its use in specific contexts, and that understanding language requires an understanding of the social and cultural practices in which it is embedded. Although large language models and diffusion models in artificial intelligence may be able to generate responses that appear to be intelligent, they may not truly understand the meaning of language as humans do.[4] This stance is closely related to the profoundly skeptical notion that Michel Foucault had in regard to authorship. Assigning authorship rather to language than to the author itself.[5] Somehow, this is a very fitting position in our contemporary age, where authorship as a concept is exposed to ever, increasing erosion.

Throughout the course of this book, we have also examined the notions of commorancies, vestures, and other stuff. By utilizing computational AI diffusion models to comprehend and analyze how people interact with their surroundings, designers can conceive of more human-focused and sustainable designs that enhance the lives of those who inhabit them. One of the key concepts we have explored is the idea of estrangement or defamiliarization, a technique used by Viktor Shklovsky[6] to make the familiar seem strange in order to prompt new ways of seeing and thinking. We have seen how this concept can be applied to architecture design, as computational AI diffusion models can prompt designers to think outside their comfort zones and create novel and innovative designs. In the neighborhood of Shklovsky's concept, we can find the territory of the uncanny, or das unheimliche, as defined by Sigmund Freud.[7] The projects in this book demonstrate how this principle can be used to explore the means by which technology and architecture can engender feelings of discomfort or unease, but also curiosity and provocation, and how designers can use this effect to create singular and unforgettable experiences for users.

The Wicked and the Tame

Architecture and engineering, like many fields, are rife with concepts that demand attention, and among these concepts, we must also consider the duality of the wicked and the tamed problem. A wicked problem, one that is complex, multifaceted, and without a definitive resolution, stands in opposition to the tamed problem, which is reducible and can be resolved by established methods and techniques. Architecture and engineering often present designers with wicked problems, those in which form, function, environmental impact, and social context all converge in a tangle of complexity, occasionally bordering on the irreducible. When confronting these wicked problems head-on, the designer's creative instincts are put to the test, as the resolution of these issues involves much more than merely choosing a single path forward. Computational AI diffusion models can aid in addressing these wicked problems by providing designers with a wealth of data and analysis to inform their decisions. Basically, interrogating the entire (available) cultural output of humanity through immense datasets. Yet, we must remember that the use of computational AI diffusion models does not necessarily eliminate the complexity of wicked problems. Rather, such models can tame some aspects of the problem and provide a framework for exploring possible solutions. The criticality of the designer's intuition and creativity cannot be overlooked in the face of computational assistance, as it is the designer's navigation of the problem's intricacies that will ultimately lead to the most innovative solutions.

Looking toward the future, it is clear that Artificial Intelligence will continue to play an increasingly important role in the field of architecture design. As computational AI diffusion models become more advanced and sophisticated, designers will have access to an unprecedented level of data and analysis, enabling them to create buildings, neighborhoods, and cities that are more responsive to the needs and desires of their users. However, as we move forward, it is important to remember the value of human creativity and intuition in the design process. By combining the power of computational AI diffusion models with human creativity and intuition, we can create a future of architecture that is both technologically advanced and deeply human-centered.

Glossary

Artificial Intelligence

Artificial intelligence (AI) refers to the development of computer systems that can perform tasks that typically require human intelligence, such as visual perception, speech recognition, decision-making, and language translation. AI systems use algorithms and machine learning techniques to learn from data and improve their performance over time.

Neural Network

A neural network is a type of machine learning model that is inspired by the structure and function of the human brain. It consists of layers of interconnected nodes, called neurons, that process and transmit information. Neural networks are trained on large datasets using algorithms that adjust the weights and biases of the connections between neurons, enabling the network to learn patterns and make predictions. They are used for tasks such as image recognition, natural language processing, and speech recognition.

Commorancy

House, place of residence. An archaic term referring to a house, place of residence, or dwelling where someone lives.

Vesture

Vesture refers to clothing or attire, especially when it is considered as a symbol of rank or status. It can also refer to the covering of a particular object or surface, such as a building or a piece of furniture. It can also refer to the act of clothing or dressing someone or something. The term is often used in a poetic or formal context, and is less commonly used in everyday language.

Ostranenie

Ostranenie is an artistic technique that involves defamiliarizing or making strange ordinary objects or experiences in order to encourage readers to view them from a new perspective. It was developed by Russian literary theorist Viktor Shklovsky.

It is a technique in which the artist presents a common object or idea in an unusual or unexpected way, with the aim of making the reader see it in a new light or from a different perspective. The term is sometimes translated into English as "estrangement" or "alienation." Ostranenie is often used in modernist and postmodernist literature, as well as in other forms of art, to challenge traditional ways of thinking and encourage critical reflection.

Estrangement

In the arts, estrangement refers to a technique in which the artist deliberately creates a sense of unfamiliarity or distance between the viewer or audience and the subject matter. This can be achieved through various means, such as the use of unconventional or abstract forms, the manipulation of perspective, or the introduction of unexpected or contradictory elements. By creating a sense of estrangement, the artist seeks to challenge the viewer's assumptions and encourage critical reflection, often with the aim of promoting social or political change.

Defamiliarization

Defamiliarization, also known as "ostranenie" or "making strange" is a technique used in the arts to present familiar objects, ideas, or experiences in a new, unexpected, or unusual way. The goal of defamiliarization is to challenge the viewer's perception and encourage critical reflection by presenting a familiar subject matter in a fresh or unfamiliar way. It is commonly used in literature, visual arts, and performance art, and is often associated with modernist and post-modernist movements. The term was coined by Russian formalist literary critic Viktor Shklovsky in the early 20th century.

Diffusion Model

Diffusion models belong to a family of Artificial Intelligence (AI) algorithms that are used for generative modeling. They are based on generative diffusion processes, which involve the recursive sampling of data from a prior distribution, followed by the diffusion of the data through multiple layers of a neural network. The generative diffusion algorithm is used to generate high-quality samples of complex data, such as images or videos, by learning the underlying probability distribution of the data. It has been used in a wide range of applications, including image and video synthesis, natural language processing, and drug discovery.

Markov Chain

A Markov chain is a mathematical model used to describe the transition between different states or events over time. It is a stochastic process that assumes that the probability of moving from one state to another depends only on the current state and not on any previous states. Markov chains are widely used in A.I. for modeling various types of systems, including natural language processing, speech recognition, and machine learning. In AI, Markov chains are often used for generating or predicting sequences of events or states based on a given set of data. For example, a Markov chain model can be used to predict the next word in a sentence based on the previous words, or to generate a new sequence of images based on a set of training data. Markov chains can also be used for reinforcement learning, where an agent learns to make decisions based on the current state and expected future rewards.

Transformers

Transformers are a type of deep learning model used in generative A.I. models such as diffusion models. In the context of diffusion algorithms, transformers are used to generate high-quality samples by learning the underlying probability distribution of the data. This is achieved by using the self-attention mechanism of transformers to capture long-range dependencies and contextual information in the input data, which is essential for modeling complex distributions such as those found in images, videos, and natural language.

The use of transformers in generative models has enabled the generation of high-quality samples that are more realistic and diverse than those generated by traditional generative models. Transformers have also facilitated the development of interactive and controllable generative models that allow users to manipulate and control the generated outputs.

Machine Learning

Machine learning is a subfield of AI that involves developing algorithms and models that enable computers to learn and make predictions or decisions based on data without being explicitly programmed. Machine learning algorithms use statistical methods to identify patterns and relationships in data, and then use these patterns to make predictions or decisions about new, unseen data.

Prompt Engineering

Prompt engineering in AI image generators is a technique that involves providing a set of instructions to a pre-trained image generator model to guide it towards generating a desired image. The prompt can include text, images, or both and is designed to elicit the desired response from the image generator. It is often used in image generation tasks such as image synthesis, style transfer, and image-to-image translation, and it can be used to control the output of the image generator by specifying certain constraints or guidelines for the generated image. Prompt engineering is an essential technique in AI image generation that enables the generation of high-quality images tailored to specific tasks and applications.

Noise

In AI image generators, noise refers to the introduction of random or unpredictable elements into the generated images. Noise can be added at various stages of the image generation process, such as the input, the latent space, or the output, to create more realistic and diverse images. Noise can help AI image generators avoid overfitting to the training data and generate unique and natural-looking images. Different types and amounts of noise can be added to the image generation process to create various effects, such as creating variations of the same image or simulating natural imperfections or distortions in the generated image. The role of noise in AI image generators is critical in producing high-quality and diverse images.

Denoising

In diffusion models, denoising is performed by progressively removing noise from the generated image over multiple steps. In each step, the noise is reduced by gradually increasing the diffusion time, which is a parameter that controls the amount of diffusion or spread of the noise. Diffusion models perform denoising by modeling the process of diffusion of the noise in the image, which can be represented as a partial differential equation. By solving this equation using numerical methods, the algorithm can estimate the amount of noise present in the image and gradually remove it by diffusion.

This process is repeated for multiple iterations until the noise is reduced to an acceptable level or eliminated completely. The denoising step in diffusion models is critical in producing high-quality and accurate images. By removing noise from the generated image, the algorithm can produce more realistic and visually appealing images, which can be used in a variety of applications, such as art, design, and entertainment.

Semantic Information

Semantic information in AI refers to the meaning or context of the data or information being processed. It represents the underlying knowledge or concepts that govern the relationships between the different entities or objects in the data. Semantic information is used in AI to improve the accuracy and efficiency of various tasks such as natural language processing, image recognition, and data analysis.

By understanding the semantic meaning of the data, AI systems can identify patterns, relationships, and insights that may not be apparent from a purely statistical or computational approach. Semantic information is often represented using ontologies, which are formal models that define the concepts, relationships, and constraints within a domain.

These ontologies can be used to annotate or tag the data, allowing AI systems to better understand the semantic meaning of the data and perform more accurate and relevant analyses.

Labeling

Labeling refers to the process of assigning a specific category or class label to each data instance in the dataset. The labels provide semantic information about the data and enable supervised learning algorithms to learn patterns and relationships between the input features and the corresponding output labels. Labeling is often done manually, with human annotators carefully examining each data instance and assigning the appropriate label based on predefined criteria or rules. This process can be time-consuming and labor-intensive, especially for large datasets with many data instances. It is often the source of racial and cultural bias in datasets, when not done with the utmost care. However, labeling is a critical step in building high-quality datasets for AI applications, as it ensures that the input data is properly categorized and annotated with meaningful labels. This enables AI systems to learn from the data and make accurate predictions or classifications on new, unseen data.

Midjourney

Midjourney is a diffusion algorithm based on image language descriptions, called "prompt." It is in its functions similar to OpenAI's DALL-E and Stable Diffusion. MidJourney is capable of generating high-quality images from textual prompts, using a combination of deep learning techniques and natural language processing. It is trained on a massive dataset of images and textual descriptions, allowing it to learn to generate images. Midjourney has a wide range of potential applications, including graphic design, advertising, and digital art. Its ability to generate novel and creative images from textual descriptions has the potential to revolutionize the creative process, allowing designers and artists to quickly generate a wide variety of ideas and concepts. It was created by the San Francisco-based independent research lab Midjourney, Inc.

DallE-2

Dall-E2 is an AI image generator developed by OpenAI. It is a successor to the original Dall-E model, which was designed to generate images from textual descriptions. The name "Dall-E2" is a reference to the surrealist artist Salvador Dali and the animated character Wall-E, highlighting the model's ability to generate unusual and imaginative images. The model has generated a wide range of images, including animals, objects, scenes, and even abstract concepts like "armchair in the shape of an avocado." In summary, Dall-E2 is an advanced image generator developed by OpenAI, capable of generating high-quality images from textual prompts using deep learning and natural language processing techniques. It has broad applications in the fields of graphic design, advertising, and digital art, and has the potential to revolutionize the creative process.

ChatGPT

ChatGPT is a large language model developed by OpenAI that uses deep learning techniques to generate human-like responses to user inputs. It is based on the Transformer architecture and has been trained on a wide range of language tasks, making it a powerful tool for a variety of applications.

AlphaGO

AlphaGo is an AI program developed by Google DeepMind to play the board game Go. It uses a combination of advanced algorithms, including deep neural networks and Monte Carlo tree search, to analyze the game board and make strategic decisions. AlphaGo made history in 2016 when it defeated the world champion, Lee Sedol, in a five-game match. This achievement demonstrated the ability of AI to master complex games that were previously thought to be too difficult for computers to play at a high level and it also demonstrated their ability to advance the field of AI research.

CLIP

CLIP (Contrastive Language-Image Pre-Training) is an AI model developed by OpenAI that uses contrastive learning to understand and generate natural language descriptions of images. It can be fine-tuned for a wide range of tasks, making it a versatile tool for computer vision and natural language processing applications. CLIP is an artificial intelligence (AI) model developed by OpenAI that can understand and generate natural language descriptions of images. It is based on a technique known as contrastive learning, which involves training the model to distinguish between positive and negative examples. CLIP is pre-trained on a large dataset of images and their associated captions, which enables it to understand the visual content of an image and the language used to describe it. This pre-training allows the model to be fine-tuned for specific tasks, such as image classification or image captioning. One of the key advantages of CLIP is that it can perform a wide range of tasks without the need for additional training or specialized models. For example, it can be used to generate captions for images, perform visual question-answering tasks, or even identify objects in images without any prior knowledge of the object or its name.

Data Mining

Data mining refers to the process of discovering patterns, trends, and insights from large datasets using statistical and machine learning techniques. It involves extracting valuable information and knowledge from large and complex datasets that may be difficult to analyze using traditional methods. Data mining typically involves preprocessing the data to remove noise and outliers, identifying relevant variables or features, selecting appropriate algorithms to analyze the data, and interpreting the results.

Data Bias

Data bias refers to the presence of systematic errors or inaccuracies in a dataset that can affect the outcomes of data analysis and machine learning algorithms. Data bias can occur due to various reasons, including sampling bias, measurement bias, or human bias in data collection, processing, or interpretation. For example, if a dataset contains more data from one demographic group than others, it may lead to biased conclusions or decisions when using that data to train a machine learning model. Similarly, if the data collection process is biased or flawed, it can lead to inaccurate conclusions and decisions. Data bias can have serious consequences, particularly in areas such as healthcare, finance, and criminal justice, where decisions based on data analysis can have a significant impact on people's lives. Therefore, it is essential to identify and address data bias in datasets used for machine learning and data analysis to ensure fair and accurate outcomes.

Dataset

A dataset is a collection of data that is organized and stored in a specific format for use in data analysis, machine learning, and other data-driven applications. A dataset can contain various types of data, such as numerical data, text data, images, audio, video, voxel, or mesh 3D models. Datasets are often used to train machine learning algorithms to make predictions or decisions based on patterns and trends found in the data. The quality and size of the dataset can significantly impact the accuracy and effectiveness of the machine learning model. Datasets can be sourced from various places, including public data repositories, private data sources, and data scraping from websites or other sources. To ensure the integrity and accuracy of a dataset, it is essential to perform data cleaning, normalization, and other preprocessing steps to eliminate errors and inconsistencies in the data.

Deep Learning

Deep learning is a subfield of machine learning that involves the use of artificial neural networks with multiple layers to learn representations of data. These neural networks are designed to mimic the structure of the human brain, with each layer of neurons processing and transforming the data in increasingly abstract ways. Deep learning algorithms are capable of learning from vast amounts of data, making them particularly useful for tasks such as image generation, speech recognition, natural language processing, and predictive analytics. These algorithms can automatically discover patterns and relationships in the data, without being explicitly programmed to do so. One of the key advantages of deep learning is its ability to learn from raw data, such as images, audio, or text, without the need for manual feature engineering. However, deep learning algorithms may require significant computational resources and large amounts of training data to achieve optimal performance.

Statistical Correlation

Statistical correlation is used to identify patterns and relationships between variables in datasets. This information is then used to train models that can predict outcomes or classify data based on these correlations. For example, in image recognition, correlations between pixel values in images and corresponding labels can be used to train deep learning models. Statistical correlation is a fundamental concept in many AI applications, including natural language processing, computer vision, and predictive analytics.

Chimera

A chimera is a term used in various contexts to describe something that is composed of different parts or elements, often resulting in a hybrid or mixed entity. In mythology, a chimera is a fire-breathing creature with the head of a lion, body of a goat, and the tail of a serpent. In genetics, a chimera refers to an organism that contains cells from two or more genetically distinct individuals, often resulting in unique characteristics. In AI and computer science, a chimera model refers to a hybrid model that combines multiple types of neural networks or other AI algorithms to achieve improved performance or solve complex problems.

Synthetic Imagination

Synthetic imagination refers to the ability to create or imagine something new or unique using a combination of existing ideas, concepts, and knowledge. It involves synthesizing different elements to form something entirely original. In the context of diffusion models, synthetic imagination refers to the ability of the model to generate novel and creative images that do not exist in reality. It involves combining existing image features and concepts in a new and imaginative way to produce visually striking and unique outputs. The AI model is trained on large datasets and uses complex algorithms to learn patterns and create new variations of images that can be used in various applications such as art, design, and marketing.

Taxonomy

Taxonomy in art history refers to the classification of artworks based on shared characteristics or attributes, such as style, period, genre, subject matter, or technique. Taxonomies can help scholars, curators, and collectors organize, compare, and analyze artworks, as well as establish connections and distinctions between different artistic traditions or movements. Taxonomies in art history can be fluid and subject to change, as new discoveries, interpretations, and perspectives may challenge previous categorizations and lead to the emergence of new taxonomies.

Contributor Profiles

Cesare Battelli

Cesare Battelli, founder of the platform visionary-architecture, is an architect dedicated to experimentation in the field of visionary art and architecture and the realization of architectural projects. He teaches and has held various workshops and served on the jury of numerous universities. He is currently a researcher at the UAH University.

Kory Bieg

Kory Bieg is an Associate Professor and Program Director for Architecture at The University of Texas at Austin. Bieg founded OTA+, an architecture, design, and research office that specializes in the development and use of current, new, and emerging digital technologies for the design and fabrication of buildings, building components, and experimental installations.

Daniel Bolojan

Daniel Bolojan is an assistant professor of AI and Computational Design at FAU, director of the Creative AI Lab, and a Ph.D. candidate at Die Angewandte Vienna. His research focuses on developing deep learning strategies for architectural design, addressing shared- agency, designer creativity, and augmentation of design potency. He developed the DeepHimmelblau Neural Network as Computational Design Specialist at Coop Himmelb(l)au, with the aim of augmenting design processes and designers' abilities.

Mario Carpo

Mario Carpo is an architectural historian and critic, currently the Reyner Banham Professor of Architectural History and Theory at the Bartlett, University College London and the Professor of Architectural Theory at the Institute of Architecture of the University of Applied Arts (die Angewandte) in Vienna. His research and publications focus on the history of early modern architecture and on the theory and criticism of contemporary design and technology. He is the author of Architecture in the Age of Printing (2001), The Alphabet and the Algorithm (2011), Beyond Digital: Design and Automation at the End of Modernity (2023), and other books.

Niccolo Casas

Niccolo Casas is an Italian architect and professor. Niccolo Casas is Principal and Founder of Niccolo Casas Architecture and a Ph.D. candidate at The Bartlett UCL. His multidisciplinary practice for research and architecture seeks to combine several fields of specialization to offer a unique perspective on the academic discipline and profession.

Virginia San Fratello

Virginia San Fratello is chair of the Department of Design at San José State University in Silicon Val ley. She is a partner in the design studio Rael San Fratello and co-founder of Emerging Objects, a 3D printing make-tank specializing in innovations for 3D printing architecture.

Soomeen Hahm

Soomeen Hahm is the founder of SoomeenHahm Design Ltd, a London-based design studio focusing on design research and tackling the issues of computational paradigms in architecture. Her latest research focuses on AR/VR, wearable machines, and humancomputer interactions. Academically she is currently a design faculty and robotic researcher at the Southern California Institute of Architecture (SCI-Arc).

Joy Knoblauch

Joy Knoblauch, Ph.D., is an associate professor of Architecture at the University of Michigan. She is the author of *The Architecture of Good Behavior, Psychology and Modern Institutional Design in Postwar America*. She is currently writing a history of resistance to technology, focusing on the history of ergonomics 194x-201x.

Daniel Koehler

Daniel Koehler, an architect and urbanist, co-founded lab-eds and is currently an assistant professor for architecture computation at UT Austin. Previously researching at UCL Bartlett and Innsbruck University, his work is featured in the Centre Pompidou's permanent collection and authored "The Mereological City." Daniel's research explores AI's impact on cities and their architecture.

Immanuel Koh

Immanuel Koh is an assistant professor at the Faculty of Design & Artificial Intelligence and Architecture & Sustainable Design at the Singapore University of Technology & Design. Trained at the Architectural Association (AA) and Zaha Hadid Architects, he holds a Ph.D. from the School of Computer Sciences and Institute of Architecture at EPFL.

Andrew Kudless

Andrew Kudless is an artist, designer, and educator based in Houston, Texas. He is the principal of Matsys and the Kendall Professor at the University of Houston Hines College of Architecture and Design. At UH, he is also the Director of the Construction Robotics and Fabrication Technologies (CRAFT) Lab.

Hanjun Kim

Hanjun Kim is an architectural designer and educator with expertise in computational and generative design. He is an associate of SoomeenHahm Design, and a course tutor at the Architectural Association Design Research Lab (AADRL) in London, where he also completed his Master of Architecture degree.

Bart Lootsma

Bart Lootsma lives in Linz, Austria, and is a historian, theoretician, critic, and curator in the fields of architecture, design, and the visual arts. He was a professor for Architectural Theory at the University of Innsbruck and guest professor at several other universities.

Elena Manferdini

Elena Manferdini is the principal of Atelier Manferdini in Los Angeles and Graduate Programs Chair at SCI-Arc. She is a leading voice in contemporary design culture and education. She received the ICON Award, a prize that recognizes iconic women who have made an indelible mark on society through their work, character, and creative leadership.

Ryan Vincent Manning

Ryan Vincent Manning is a Texas-based designer investigating new building models that look to incorporate ideas from the AI, computation, and fashion industry to produce buildings and objects into soft familiarities. He is the principal of quirkd33, a research and development collaborative in products, writings, and architectural ideas.

Sandra Manninger

Sandra Manninger is an architect, researcher, and educator. Born and educated in Austria, she co-founded SPAN Architecture with Matias del Campo in 2003. Her award-winning projects have been published and exhibited internationally, including at La Biennale di Venezia, MAK, and Autodesk Pier 1, and have been included in the permanent collections of the FRAC, Design Museum in Munich, and the Albertina in Vienna. Sandra Manninger has taught internationally at the TU Vienna, University for Applied Arts, DIA Bauhaus in Dessau, UPenn, Tongji and Tsinghua Universities, the University of Michigan, and the Royal Melbourne Institute of Architecture. She currently serves as an associate professor at the New York Institute of Technology (NYIT)

Lev Manovich

Lev Manovich is an artist, writer, and one of the world's most influential digital culture theorists. He was included in the lists of "25 People Shaping the Future of Design" (Complex, 2013) and "50 Most Interesting People Building the Future" (Verge, 2014). Manovich is a presidential professor in Data Science and Computer Science at the Graduate Center, City University of New York, and a director of the Cultural Analytics Lab. Manovich published 180 articles and 15 books that include *AI Aesthetics*, *Cultural Analytics*, *Instagram and Contemporary Image*, and *The Language of New Media* described as "the most suggestive and broad-ranging media history since Marshall McLuhan."

Asma Mehan

Dr. Asma Mehan, assistant professor at Texas Tech University's Huckabee College of Architecture, specializes in architectural humanities and critical urban studies. She authored "Tehran: from Sacred to Radical" (2022) and "Kuala Lumpur: Community, Infrastructure, and Urban Inclusivity" (2020), both published by Routledge.

Sina Mostafavi

Dr. Sina Mostafavi is an associate professor at Texas Tech University's Huckabee College of Architecture, focusing on innovative applications of emerging materials and technologies. He is the founder of the award-winning SETUParchitecture studio. Holding a doctoral degree from TU Delft, Mostafavi has authored the book *Hybrid Intelligence in Architectural Robotic Materialization.*

Rasa Navasaityte

Rasa Navasaityte, an architect and lab-eds co-founder, lectures at UT Austin, specializing in urban ecologies, housing, and computation. With experience at UCL Bartlett and Innsbruck University, her work has been widely published and exhibited. Her Ph.D., "Between the Parts," explores an architectural framework for ecological urban form.

Alicia Nahmad Vazquez

Alicia Nahmad Vazquez is an architect and entrepreneur. She is the founder of The Circular Factory (CF) and works as an associate professor at the University of Calgary, where she codirects the Laboratory for Integrative Design. Previously, Alicia worked as a studio master at the Architectural Association DRL and developed design tools for practices like Populous and Zaha Hadid Architects. Her research has been widely published internationally.

Igor Pantic

Igor Pantic is a London-based architect and designer and a lecturer at the UCL Bartlett School of Architecture. His work focuses on computational design and application of Mixed Reality in architecture, exploring the ways in which technology influences how we design, make, and perceive our environment. Prior to establishing his own design practice, Igor spent a number of years working for Zaha Hadid Architects in London.

Kyle Steinfeld

Kyle Steinfeld makes, writes, and teaches about computational design as a cultural practice. As an associate professor at UC Berkeley, he applies techniques drawn from artificial intelligence to architectural design, and aims to reveal the overlooked capacities of computation through creative work, writing, and speculative tool-making.

Marco Vanucci

Marco Vanucci is an architect and the design director of Opensystems Architecture. He designed and built projects in Europe, Middle East, and China. He's a lecturer in architectural design at London South Bank University. He also taught at the AA, KTH Stockholm, and Cardiff University. His research focuses on the origin of parametric, associative design methods and, lately, on the relationship between architecture and AI.

Dustin White

Dustin White is a New York-based designer and educator. He is the owner of Dustin White Design, an interdisciplinary design research office, which operates across multiple scales, from furniture to buildings, and employs hybrid practices merging traditional, computational, and emerging technologies to explore opportunities for a reinvigorated understanding of architectural materials and spatial conditions.

References

Prologue

1: Forlano, Laura. (2017). Posthumanism and design. She Ji: The Journal of Design, Economics, and Innovation 3(1):16–29. https://doi.org/10.1016/J.SHEJI.2017.08.001.

2: Boden, Margaret A. (2007). Creativity in a nutshell. Think 5 (15):83–96.

3: McGuigan, Cathleen. (2010). The Death of Starchitecture. Newsweek. https://www.newsweek.com/death-starchitecture-73049 retrieved Apr 3rd 2023.

4: Carpo, Mario. (2022). Neural Architecture – Design and Artificial Intelligence, Matias del Campo. ORO Editions, pp. 14.

5: Rorty, Richard. (1967). The Linguistic Turn: Recent Essays in Philosophical Method. Chicago: University of Chicago Press.

6: Rorty, Richard. (2007). Wittgenstein and the Linguistic Turn In Cultures. Conflict – Analysis-Dialogue: Proceedings of the 29th International Ludwig Wittgenstein-Symposium in Kirchberg, Austria edited by Christian Kanzian and Edmund Runggaldier, 3–20. Berlin, Boston: De Gruyter, 2007. https://doi.org/10.1515/9783110328936.3.

7: Lee, Benjamin. (1997). Introduction In Talking Heads: Language, Metalanguage, and the Semiotics of Subjectivity, 1–15. New York, USA: Duke University Press. https://doi.org/10.1515/9780822382461-002.

8: Silverman, Hugh J. (1979). Merleau-Ponty on Language and Communication (1947–1948). Research in Phenomenology 9:168–181. http://www.jstor.org/stable/24654334.

9: Timcke, Scott. (2019). Foucault, White, and the Linguistic Turn in Western Historiography. History in Action 6(1):1–12, Forthcoming, Available at SSRN: https://ssrn.com/abstract=3317621.

10: Michels, James. (1995). ROLAND BARTHES: AGAINST LANGUAGE. ETC: A Review of General Semantics 52(2):155–173. http://www.jstor.org/stable/42577631.

11: Casey, Edward S. (1997). The Fate of Place: A Philosophical History. 1st ed. University of California Press. http://www.jstor.org/stable/10.1525/j.ctt2jcbw8.

12: Pepper, Stephen C. (1950). Some comments on Professor Kahn's paper. The Journal of Aesthetics and Art Criticism 9(1):51–55. Accessed April 5, 2023. https://doi.org/10.2307/426102.

13: James, F. R. (1931). Book review: vestments and vesture: a manual of liturgical art. The Downside Review 49(2):377–380. https://doi.org/10.1177/001258063104900223.

14: Barthes Roland and Andy Stafford. (2006). The Language of Fashion English ed. Oxford: Berg.

Suppositions

1: Lugton, Robert C. (1965). Ludwig wittgenstein: the logic of language. ETC: A Review of General Semantics 22(2):165–192. http://www.jstor.org/stable/42574113.

2: Foucault, Michel. (1982). The subject and power. Critical Inquiry 8(4) 777–195. http://www.jstor.org/stable/1343197.

3: Maxwell, Nicholas. (2017). Karl Popper, Science and Enlightenment. UCL Press.
https://doi.org/10.2307/j.ctt1vxm8p6.

4: Gatys, Leon A., Ecker, Alexander S., and Bethge, Matthias. (2015). A Neural Algorithm of Artistic Style. ArXiv: 1508.06576 (August–September 2015); Leon A. Gatys, Alexander S. Ecker, and Matthias Bethge. (2016). "Image Style Transfer Using Convolutional Neural Networks," 2016 IEEE Conference on Computer Vision and Pattern Recognition (June 2016): 2414–2423.

5: Reed, Scott. et al. (2022). A Generalist Agent Transactions on Machine Learning Research, 11/2022, arXiv:2205.06175

6: Baudrillard, Jean. (1994). Simulacra and Simulation. Ann Arbor: The University of Michigan Press. p.2.

7: Kissick, Dean. (2013). Didn't I see you on the cover of i-D?, i-D 326, Pre-Fall 2013, The Street Issue.

8: https://www.archdaily.com/content/about?ad_source=jv-header&ad_name=hamburger_menu, Accessed 20230317.

9: Jacob, Sam. (2017). Architecture Enters the Age of Post-Digital Drawing, Metropolis. http://www.metropolismag.com/architecture/architecture-enters-age-post-digital-drawing/, accessed 20170716.

10: Jacob, Sam. (2017). Architecture Enters the Age of Post-Digital Drawing, Metropolis. http://www.metropolismag.com/architecture/architecture-enters-age-post-digital-drawing/, accessed 20170716.

11: Davide, Tommaso Ferrando, Bart, Lootsma, and Kanokwan, Trakulyingcharoen. (2020). Italian Collage, Lettera Ventidue, Siracusa.

12: Bridle, James. (2023). The stupidity of AI, The Guardian, 20230316. https://www.theguardian.com/technology/2023/mar/16/the-stupidity-of-ai-artificial-intelligence-dall-e-chatgpt?fbclid=IwAR3uIea7PVtxFIwqU-bTK8guAWrVJTpD6WbiByn8qo5wy_KK14k88Hnn1Ns, accessed 20230317.

13: Michels, Karen. (1989). Der Sinn der Unordnung. Arbeitsformen im Atelier Le Corbusier, Vieweg, Braunschweig/Wiesbaden: Vieweg.

14: Carpo, Mario. (2017). The Second Digital Turn, Design Beyond Intelligence. MIT Press. pp. 102–103.

15: Idem.

16: Baudrillard, Jean. (1994). Simulacra and Simulation,. Ann Arbor: The University of Michigan Press. p.1.

17: Baudrillard, Jean. Idem. p.6.

18: Barthes, Roland. (2002). Marsmannetjes, Mythologieën. Utrecht: UItgeverij IJzer. pp. 43–45.

19: Eco, Umberto. (1990). Die Struktur des schlechten Geschmacks, in: Im Labyrinth der Vernunft, Texte über Kunst und Zeichen. Leipzig: Reklam. p. 246.

20: Sontag, Susan. (1966). "Notes on Camp." in: Against Interpretation and Other Essays. New York: Farrar, Straus & Giroux. p. 279.

21: Kissick, Dean. Baudrillard, Jean. (1994). Simulacra and Simulation. Ann Arbor: The University of Michigan Press.

22: Zwier, Karen. (2018). Methodology in Aristotle's theory of spontaneous generation. Journal of the History of Biology 51(2):355–386.

23: Tubaro, P., Casilli, A. A., and Coville, M. (2020). The trainer, the verifier, the imitator: Three ways in which human platform workers support artificial intelligence. Big Data & Society 7(1). https://doi.org/10.1177/2053951720919776.

24: Kelner, D. (2022). The State of AI 2019: Divergence. https://www.stateofai2019.com/introduction retrieved Nov. 15th 2022.

25: Perrigo, B. (2022). Inside Facebook's African Sweatshop, time.com. https://time.com/6147458/facebook-africa-content-moderation-employee-treatment/ retrieved Nov.15th 2022.

26: Newman, A. (2022). I Found Work on an Amazon Website. I Made 97 Cents an Hour. New York Times. https://www.nytimes.com/interactive/2019/11/15/nyregion/amazon-mechanical-turk.html, retrieved Nov. 6th 2022.

27: Redmon, Joseph, Divvala, Santosh Kumar, Girshick, Ross B. and Farhadi, Ali. (2016). "You only look once: Unified, real-time object detection." 2016 IEEE Conference on Computer Vision and Pattern Recognition (CVPR) (2016): 779–788.

28: Xin, M. and Wang, Y. (2019). Research on image classification model based on deep convolution neural network. Journal of Image and Video Processing. 40. https://doi.org/10.1186/s13640-019-0417-8.

29: Borad, A. (2022). Understanding Object Localization with Deep Learning, einfochips.com. https://www.einfochips.com/blog/understanding-object-localization-with-deep-learning/ retrieved Nov. 6th 2022.

30: In 1985 the Psychologist and Cognitive Scientist George A. Miller started working with his team at Princeton on WordNet, a lexical database for the English language. Historyofinformation.com https://www.historyofinformation.com/detail.php?entryid=2471 retrieved Nov. 18th 2022.

31: Ngo, Vuong M., Cao, Tru Hoang, and Le, Tuan M. V. (2018). 'WordNet-Based Information Retrieval Using Common Hypernyms and Combined Features. ArXiv abs/1807.05574.

32: Elberrichi, Zakaria, Rahmoun, Abdellatif, and Bentaallah, Mohamed. (2008). Using WordNet for text categorization. The International Arab Journal of Information Technology 5:16–24.

33: Pal, A. R. and Saha, D. (2014). "An approach to automatic text summarization using WordNet," 2014 IEEE International Advance Computing Conference (IACC). 2014, pp. 1169–1173. https://doi.org/10.1109/IAdCC.2014.6779492.

34: Saija, Krunal, Sangeetha, S., and Shah, Viral. (2019). "WordNet-Based Sign Language Machine Translation: From English Voice to ISL Gloss." IEEE 16th India Council International Conference (INDICON). pp. 1–4.

35: Rigutini, Leonardo, Diligenti, Michelangelo, Maggini, Marco, and Gori, Marco. (2012). Automatic generation of crossword puzzles. International Journal on Artificial Intelligence Tools 21(03):1250014.

36: Gershgorn, D. (2022). The data that transformed AI research—and possibly the world, QUARTZ. https://qz.com/1034972/the-data-that-changed-the-direction-of-ai-research-and-possibly-the-world retrieved Nov. 16th 2022.

37: Harrison, A. (2023). Princeton University's report on diversity details university efforts, racial growth of student population. https://centraljersey.com/2021/10/26/princeton-universitys-report-on-diversity-details-university-efforts-racial-growth-of-student-population/ retrieved Feb. 12th 2023.

38: Denton, E., Hanna, A., Amironesei, R., Smart, A., and Nicole, H. (2021). On the genealogy of machine learning datasets: A Critical History of ImageNet. Big Data & Society 8(2). https://doi.org/10.1177/20539517211035955.

39: Gershgorn, D. (2022). The data that transformed AI research – and possibly the world, QUARTZ. https://qz.com/1034972/the-data-that-changed-the-direction-of-ai-research – and possibly the world retrieved Nov. 16th 2022.

40: Gershgorn, D. (2022). The data that transformed AI research – and possibly the world, QUARTZ. https://qz.com/1034972/the-data-that-changed-the-direction-of-ai-researchandpossiblytheworld retrieved Nov. 16th 2022.

41: Gerald, M. L. (2000). The Turk Chess Automaton. McFarland & Co Inc Pub.

42: Gershgorn, D. (2022). The data that transformed AI research – and possibly the world, QUARTZ. https://qz.com/1034972/the-data-that-changed-the-direction-of-ai-research and possibly the world retrieved Nov. 16th 2022.

43: Buhrmester, M., Kwang, T., and Gosling, S. D. (2011). Amazon's Mechanical Turk: a new source of inexpensive, yet high-quality, data? Perspectives on Psychological Science 6(1):3–5. https://doi.org/10.1177/1745691610393980.

44: Dholakia, U. (2017). Just how many Amazon MTurk Survey-Takers are there? Psychology Today. https://www.psychologytoday.com/us/blog/the-science-behind-behavior/201701/just-how-many-amazon-mturk-survey-takers-are-there retrieved Nov. 17th 2022.

45: Difallah, Djellel, Filatova, Elena, and Ipeirotis, Panos. (2018). "Demographics and Dynamics of Mechanical Turk Workers." In: WSDM '18: Proceedings of the Eleventh ACM International Conference on Web Search and Data Mining. 135–143. https://doi.org/10.1145/3159652.3159661.

46: Gershgorn, D. (2022). The data that transformed AI research – and possibly the world, QUARTZ. https://qz.com/1034972/the-data-that-changed-the-direction-of-ai-researchandpossiblytheworld retrieved Nov. 16th 2022.

47: Unfortunately, the website of Techlist depicting this map does not exist anymore, but you still can find it here. http://turktools.net/images/techlist.png.

48: Toonders, J. (2023). Data is the New Oil of the Digital Economy. https://www.wired.com/insights/2014/07/data-new-oil-digital-economy/ retrieved Feb. 12th 2023.

49: See Bridle, J. (2018). New Dark Age : Technology and the End of the Future. London: Verso.

50: Denton, E., Hanna, A., Amironesei, R., Smart, A. and Nicole, H. (2021). On the genealogy of machine learning datasets: A critical history of ImageNet. Big Data & Society 8(2). https://doi.org/10.1177/20539517211035955.

51: Denton, E., Hanna, A., Amironesei, R., Smart, A. and Nicole, H. (2021). On the genealogy of machine learning datasets: A Critical History of ImageNet. Big Data & Society 8(2). https://doi.org/10.1177/20539517211035955.

52: Russakovsky, O., Deng, J. Su, H. et al. (2015). Imagenet large-scale visual recognition challenge. International Journal of Computer Vision 115(3): 211–252.

53: An ongoing project at the Ar2IL laboratory at the Taubman School of Architecture and Urban Planning. University of Michigan.

54: Crawford, Kate and Paglen, Trevor. (2019). Excavating AI: The Politics of Training Sets for Machine Learning (September 19, 2019) https://excavating.ai.

55: I highly recommend reading Kate Crawford's "Atlas of AI" on the history of dataset creation and the reasons why we find so much bias in current datasets.

56: Plasek, A. (2016). "On the cruelty of really writing a history of machine learning." IEEE Annals of the History of Computing 38:6–8. https://doi.org/10.1109/MAHC.2016.43.

57: Foucault, M. (1977). 1926–1984. Discipline and Punish: the Birth of the Prison. New York : Pantheon Books.

58: Koopman, C. (2019). How We Became Our Data: A Genealogy of the Informational Person. Chicago, IL, USA: University of Chicago Press, 2019

59: Hatam, Sara, Muwafaq Al-Ghabra, Iman, and Ghabra, Al. (2021). Barthes' semiotic theory and interpretation of signs. International Journal of Research in Social Sciences and Humanities 11:470–482. https://doi.org/10.37648/ijrssh.v11i03.027.

60: Denton, E., Hanna, A., Amironesei, R., Smart, A. and Nicole, H. (2021). On the genealogy of machine learning datasets: A critical history of ImageNet. Big Data & Society 8(2). https://doi.org/10.1177/20539517211035955.

61: Reizen, P. Z. (2018). Happiness for Humans. New York: Grand Central Publishing.

62: Onuoha, Mimi and Nuncera, Diana. (2018). A People's Guide to AI. Detroit: A People's Guide to Tech.

63: Kate Crawford estimates that running one Natural Language Processing model is as energy intensive as flying from New York to Beijing 125 times producing around 660,000 pounds of carbon dioxide. Kate Crawford, Atlas of AI: Power, Politics, and the Planetary Costs of Artificial Intelligence. (New Haven: Yale University Press, 2021). P. 42.

64: Harry, Braverman. (1974). Labor and Monopoly Capital. the Degradation of Work in the Twentieth Century. New York: Monthly Review Press.

65: Zuboff, Shoshana. (2020). In the Age of the Smart Machine the Future of Work and Power. New York, NY: Basic Books, 1989; Shoshana Zuboff (2020). The Age of Surveillance Capitalism: The Fight for a Human Future at the New Frontier of Power.

66: https://blackinai.github.io/#/

67: Michelle, Gwynne and Midjourney. (2023). "Renaissance Tracksuits," on Midjourney Official Facebook page February 25, 2023.

68: Robin Evans. (2000). The Projective Cast, Architecture and Its Three Geometries page xxviii.

69: Yang, Ling, Zhang, Zhilong, Hong, Shenda, Xu, Runsheng, Zhao, Yue, Shao, Yingxia, Zhang, Wentao, Yang, Ming-Hsuan, and Cui, Bin. (2022). Diffusion Models: A Comprehensive Survey of Methods and Applications. ArXiv abs/2209.00796.

70: Goodfellow, I. J., Mirza, M., Xu, B., Ozair, S., Courville, A., and Bengio, Y. (2014). Generative Adversarial Networks. arXiv. https://doi.org/10.48550/arXiv.1406.2661.

71: Reed, Scott E., Akata, Zeynep, Yan, Xinchen, Logeswaran, Lajanugen, Schiele, Bernt, and Lee, Honglak. (2016). Generative Adversarial Text to Image Synthesis. ArXiv abs/1605.05396.

72: Forsyth, D. A. and Ponce, Jean. (2002). Computer Vision: A Modern Approach. Prentice Hall Professional Technical Reference.

73: Manning, Christopher D. and Schütze, Hinrich. (1999). Foundations of Statistical Natural Language Processing. Cambridge, MA: The MIT Press.

74: Shafique, Muhammad Akmal, Naseer, Mahum, Theocharides, Theocharis, Kyrkou, C., Mutlu, Onur, Orosa, Lois, and Choi, Jungwook. (2020). Robust machine learning systems: challenges, current trends, perspectives, and the road ahead. IEEE Design & Test 37: 30–57.

75: Bilos, Marin, Rasul, Kashif, Schneider, Anderson L., Nevmyvaka, Yuriy, and Günnemann, Stephan. (2022). Modeling Temporal Data as Continuous Functions with Process Diffusion. ArXiv abs/2211.02590.

76: Ahishakiye, Emmanuel, Van Gijzen, Martin Bastiaan, Tumwiine, Julius, Wario, Ruth, and Obungoloch, Johnes. (2021). A survey on deep learning in medical image reconstruction, Intelligent Medicine 1(3):118–127, ISSN 2667-1026, https://doi.org/10.1016/j.imed.2021.03.003.

77: Keith, John A., Vassilev-Galindo, Valentin, Cheng, Bingqing, Chmiela, Stefan, Gastegger, Michael, Müller, Klaus-Robert, and Tkatchenko, Alexandre. (2021). Combining machine learning and computational chemistry for predictive insights into chemical systems. Chemical Reviews 121(16):9816–9872. https://doi.org/10.1021/acs.chemrev.1c00107.

78: SPAN (Matias del Campo, Sandra Manninger), Immanuel Koh, Daniel Bolojan et al.

79: Radford, Alec, Wu, Jeff, Child, Rewon, Luan, David, Amodei, Dario, and Sutskever, Ilya. (2019). Language Models are Unsupervised Multitask Learners.

80: Anokhin, Ivan, Demochkin, Kirill V., Khakhulin, Taras, Sterkin, Gleb, Lempitsky, Victor S., and Korzhenkov, Denis. (2020). "Image Generators with Conditionally-Independent Pixel Synthesis." 2021 IEEE/CVF Conference on Computer Vision and Pattern Recognition (CVPR) (2020): 14273–14282.

81: LAION. (2022). In Wikipedia. https://en.wikipedia.org/wiki/LAION

82: Xu, Tao, Zhang, Pengchuan, Huang, Qiuyuan, Zhang, Han, Gan, Zhe, Huang, Xiaolei, and He, Xiaodong. (2017). "AttnGAN: Fine-Grained Text to Image Generation with Attentional Generative Adversarial Networks." 2018 IEEE/CVF Conference on Computer Vision and Pattern Recognition (2017): 1316–1324.

83: The transformation of the 2D pixel image to the 3D model occurred by turning the pixel image into color patches in Grasshopper3D and then extruding the color patches along the z-axis. Randomly moving them in the Z direction.

84: Lin, Tsung-Yi, Maire, Michael, Belongie, Serge J., Hays, James, Perona, Pietro, Ramanan, Deva, Dollár,

Piotr and Zitnick, C. Lawrence. (2014). "Microsoft COCO: Common Objects in Context." European Conference on Computer Vision.

85: Vinyals, Oriol, Toshev, Alexander, Bengio, Samy, and Erhan, D. (2014). "Show and tell: A neural image caption generator." 2015 IEEE Conference on Computer Vision and Pattern Recognition (CVPR) (2014): 3156–3164.

86: Deng, J., Dong, W., Socher, R., Li, L. -J., Li, Kai, and Fei-Fei, Li. (2009). "ImageNet: A large-scale hierarchical image database." 2009 IEEE Conference on Computer Vision and Pattern Recognition, pp. 248–255. https://doi.org/10.1109/CVPR.2009.5206848.

87: Mansimov, Elman, Parisotto, Emilio, Ba, Jimmy, and Salakhutdinov, Ruslan. (2016). "Generating Images from Captions with Attention." CoRR abs/1511.02793 (2016): n. pag.

88: Sohl-Dickstein, Jascha Narain, Weiss, Eric A., Maheswaranathan, Niru, and Ganguli, Surya. (2015). "Deep Unsupervised Learning using Nonequilibrium Thermodynamics." ArXiv abs/1503.03585 (2015): n. pag.

89: Ley, Christophe, Babić, Slađana, and Craens, Domien. (2021). Flexible models for complex data with applications. Annual Review of Statistics and Its Application. 8. https://doi.org/10.1146/annurev-statistics-040720-025210.

90: Bahri, Yasaman, Kadmon, Jonathan, Pennington, Jeffrey, Schoenholz, Sam S., Sohl-Dickstein, Jascha, and Ganguli, Surya. (2020). Statistical mechanics of deep learning. Annual Review of Condensed Matter Physics 11(1):501–528

91: Song, Yang, Sohl-Dickstein, Jascha Narain, Kingma, Diederik P., Kumar, Abhishek, Ermon, Stefano, and Poole, Ben. (2020). "Score-Based Generative Modeling through Stochastic Differential Equations." ArXiv abs/2011.13456 (2020): n. pag.

92: Ho, Jonathan, Jain, Ajay, and Abbeel, P. (2020). "Denoising Diffusion Probabilistic Models." ArXiv abs/2006.11239 (2020): n. pag.

93: Vahdat, Arash, Kreis, Karsten, and Kautz, Jan. (2021). "Score-based Generative Modeling in Latent Space." Neural Information Processing Systems (2021).

94: Yang, Ling, Zhang, Zhilong, Hong, Shenda, Xu, Runsheng, Zhao, Yue, Shao, Yingxia, Zhang, Wentao, Yang, Ming-Hsuan, and Cui, Bin. (2022). "Diffusion Models: A Comprehensive Survey of Methods and Applications." ArXiv abs/2209.00796 (2022): n. pag.

95: Smith, Samuel L. and Le, Quoc V. (2017). "A Bayesian Perspective on Generalization and Stochastic Gradient Descent." ArXiv abs/1710.06451 (2017): n. pag.

96: Baum, L. E. and Petrie, T. (1966). Statistical inference for probabilistic functions of finite state Markov chains. The Annals of Mathematical Statistics 37(6): 1554–1563.

97: Strauss, D. (2022). Skills and tools: A philosophical perspective on technology. https://www.academia.edu/37037818/Skills_and_tools_A_philosophical_perspective_on_technology retrieved Nov 23rd 2022.

98: del Campo, M. and Manninger, Sandra. (2011). Artifact & Affect. PARC Pesquisa em Arquitetura e Construção. 2. 12. 10.20396/parc.v2i7.8634580.

99: Biro, D., Haslam, M., and Rutz, C. (2013). Tool use as adaptation. Philosophical Transactions of the Royal Society B: Biological Sciences 368(1630):20120408. https://doi.org/10.1098/rstb.2012.0408.PMID: 24101619; PMCID: PMC4027410.

100: Strauss, D. (2022). Skills and tools: A philosophical perspective on technology. https://www.academia.edu/37037818/Skills_and_tools_A_philosophical_perspective_on_technology retrieved Nov 23rd 2022.

101: Hagen, Joel B. (2009). Descended from Darwin? George Gaylord Simpson, Morris Goodman, and primate systematics. Transactions of the American Philosophical Society 99(1):93–109. http://www.jstor.org/stable/27757426.

102: Parry, Richard. (2021). "Episteme and Techne", The Stanford Encyclopedia of Philosophy (Winter 2021 Edition), Edward N. Zalta (ed.). https://plato.stanford.edu/archives/win2021/entries/episteme-techne/> retrieved Nov 23rd 2022.

103: Wang, Mei and Deng, Weihong. (2021). Deep Face Recognition: A Survey. Neurocomputing 429:215–244.

104: Tiwari, Pooja, Mehta, Simran, Sakhuja, Nishtha, Kumar, Jitendra, and Singh, Ashutosh Kumar. (2021). "Credit Card Fraud Detection using Machine Learning: A Study." ArXiv abs/2108.10005 (2021): n. pag.

105: Kumar, Y., Koul, A., Singla, R., and Ijaz, M. F. (2022). Artificial intelligence in disease diagnosis: A systematic literature review, synthesizing framework and future research agenda. Journal of Ambient Intelligence and Humanized Computing 13:1–28. https://doi.org/10.1007/s12652-021-03612-z. Epub ahead of print. PMID: 35039756; PMCID: PMC8754556.

106: Nassif, A. B., Shahin, I., Attili, I., Azzeh, M., and Shaalan, K. (2019). Speech recognition using deep neural networks: A systematic review. IEEE Access 7:19143–19165. https://doi.org/10.1109/ACCESS.2019.2896880.

107: Gamarra, Walter, Martínez, Elvia, Cikel, Kevin, Santacruz, Maira, Arzamendia, Mario, Gregor, Derlis, Villagra, Marcos, and Colbes, José. (2021). "Deep Learning for Traffic Prediction with an Application to Traffic Lights Optimization." 2021 1st International Conference on Artificial Intelligence and Data Analytics (CAIDA), pp. 31–36, https://doi.org/10.1109/CAIDA51941.2021.9425158.

108: Tsolaki, Kalliopi, Vafeiadis, Thanasis, Nizamis, Alexandros, Ioannidis, Dimosthenis, and Tzovaras, Dimitrios. (2022). Utilizing machine learning on freight transportation and logistics applications: A review. ICT Express, ISSN 2405-9595, https://doi.org/10.1016/j.icte.2022.02.001.

109: Nooteboom, Bart. (2020). Objects, relations, potential and change. Open Philosophy 3(1):53–67. https://doi.org/10.1515/opphil-2020-0004.

110: Sommer, Ulrike. (2010). "Anthropology, Ethnography and Prehistory – A Hidden Thread in the History of German Archaeology." In: Domanska, L. Grøn, O. Hardy, K. (Eds.) Archaeological Invisibility and Forgotten Knowledge. Proceedings of a Conference 5 – 8 Sept 2007, University of Lodz. BAR Int. 2183, Oxford: Archaeopress 2010, 6–22. 2183 (2010).

111: 'Prologue: Machine Civilization and the Transformation of the Earth'Dark Skies: Space Expansionism, Planetary Geopolitics, and the Ends of Humanity, New York, 2020; online edn, Oxford Academic. 19 Mar. 2020 https://doi.org/10.1093/oso/9780190903343.002.0008, accessed 13 Dec. 2022.

112: Trinkaus, Charles. (1949). The problem of free will in the renaissance and the reformation. Journal of the History of Ideas 10(1):51–62. https://doi.org/10.2307/2707199.

113: Zalta, Edward N. (2022). "Gottlob Frege", The Stanford Encyclopedia of Philosophy (Fall 2022 Edition), Edward N. Zalta & Uri Nodelman (eds.), URL = <https://plato.stanford.edu/archives/fall2022/entries/frege/>.

114: Churchill, John. (1984). Wittgenstein on the phenomena of belief. International Journal for Philosophy of Religion 16(2):139–152. http://www.jstor.org/stable/40012635.

115: Landesman, Charles. (1976). Locke's theory of meaning. Journal of the History of Philosophy 14(1):23–35. https://doi.org/10.1353/hph.2008.0165.

Commorancies

1: Casey, Edward S. (1997). The Fate of Place: A Philosophical History. 1st ed. University of California Press http://www.jstor.org/stable/10.1525/j.ctt2jcbw8.

2: Malpas, Jeff. (2012). Heidegger and the Thinking of Place: Explorations in the Topology of Being. The MIT Press. http://www.jstor.org/stable/j.ctt5vjp35.

3: Casey, Edward S. (1997). The Fate of Place: A Philosophical History. 1st ed. University of California Press. http://www.jstor.org/stable/10.1525/j.ctt2jcbw8.

4: Fernand, Braudel. (1986). Civilization and Capitalism 15th–18th Century. New York: Harper & Row.

5: Foucault, Michel and Miskowiec, Jay. (1986). Of other spaces. Diacritics 16(1):22–27. https://doi.org/10.2307/464648.

6: Berry, Wendell. (2001). A Place on Earth, Counterpoint, 2001

7: Wirth, Jason M. (2019). Mountains, rivers, and the great earth: Reading Gary Snyder and Dogen in an age of ecological crisis. Journal of Beat Studies 7:88+. Gale Academic OneFile (accessed April 9, 2023).

8: McMahon, Laura. (2011). Jean-Luc Nancy and the spacing of the world. Contemporary French and Francophone Studies 15(5):623–631, https://doi.org/10.1080/17409292.2011.647877

9: Elden, Stuart. (2007). There is a politics of space because space is political: Henri Lefebvre and the production of space. Radical Philosophy Review 10:101–116. 10.5840/radphilrev20071022.

10: Wheeler, Andrea. (2023). Space: Notes on the Thought of Luce Irigaray, (accessed April 9, 2023). https://criticallegalthinking.com/2015/11/23/space-notes-on-the-thought-of-luce-irigaray/

11: Bachelard, Gaston. (2014). The Poetics of Space. London, England: Penguin Classics.

12: Raphael, Melissa. (1997). 'Introduction', Rudolf Otto and the Concept of Holiness (Oxford, 1997; online edn, Oxford Academic, 3 Oct. 2011), https://doi.org/10.1093/acprof:oso/9780198269328.003.0010, accessed 9 Apr. 2023.

13: Crampton, Jeremy W. and Elden, Stuart. (2007). "Space, Knowledge and Power: Foucault and Geography."

14: Yi-fu, Tuan. (1977). Space and Place : The Perspective of Experience. Minneapolis: University of Minnesota Press.

15: Soja, E. W. (1989). Postmodern Geographies: The Reassertion of Space in Critical Social Theory, Verso.

16: Buttimer, A. and Seamon, D. (1980). The Human Experience of Space and Place (1st ed.). Routledge. https://doi.org/10.4324/9781315684192

17: Relph, E. (2007). Spirit of Place and Sense of Place in Virtual Realities, Techné: Research in Philosophy and Technology, Philosophy Documentation Center pp 17–15

18: Malpas, Jeff. (2012). 'Heidegger in Benjamin's City', Heidegger and the Thinking of Place: Explorations in the Topology of Being (Cambridge, MA, 2012; online edn, MIT Press Scholarship Online, 22 Aug. 2013), https://doi.org/10.7551/mitpress/9780262016841.003.0012, accessed 9 Apr. 2023.

19: Arendt, Hannah. (2018). The Human Condition. 2nd ed. Chicago, IL: University of Chicago Press.

20: Mumford, L. (1961). The City in History, Harcourt, Brace and World.

21: Deleuze, Gilles and Guattari, Félix. (1986). Nomadology : The War Machine. New York NY USA: Semiotext(e).

22: Derrida, Jacques, and Hanel, Hilary P. (1990). "A Letter to Peter Eisenman." Assemblage (12):7–13. https://doi.org/10.2307/3171113.

23: Eisenman, P. (2017). Peter Eisenman: In dialogue with architects and philosopher (Vladan Djokić and Petar Bojanić (eds.)), Mimesis International.

24: Tschumi, B. (1994). Event-Cities : Praxis. Cambridge Mass: MIT Press.

25: The photograph, "View from the Window at le Gras," is now located at The University of Texas at Austin's Harry Ransom Center, about 200 feet from my office where I am using ChatGPT to help with my research for this text.

26: Benjamin, Walter. (2008). The Work of Art in the Age of Mechanical Reproduction. Translated by J.A. Underwood, Penguin Books, 24.

27: See my essay "Artificial intelligence can now make convincing images of buildings. Is that a good thing?" in the July/August 2022 issue of Architect's Newspaper for more thoughts on text-to-image AI and sketching.

28: Rosalind Krauss. (1970). "Originality and Presence," Artforum, June 1970.

29: Moussavi, Farshid. (2009). The Function of Form (ACTAR, Harvard Graduate School of Design, 2009), 19. I won't get into the legacy of difference as a design agent here, but there is a rich history of using difference as a design generator in architecture and other design disciplines. See the writing of Deleuze and Guattari, Derrida, and other post-modernists, or the architectural work of Bernard Tschumi and other deconstructivists.

30: Kristeva, Julia. (1982). Desire in Language: A Semiotic Approach to Literature and Art. Columbia University Press, 66.

31: Kristeva, Julia. (1982). Desire in Language: A Semiotic Approach to Literature and Art (Columbia University Press), 148.

32: Lumifoil was designed by Kory Bieg and Clay Odom, 2016.

33: Plume was designed and built by Kory Bieg and Clay Odom, 2022.

34: It's beyond the scope of this essay, but these AI tools fully align with much of the postmodern theories that were popular in certain architectural circles in the '90s and '00s, namely the interest in heterogeneities, multiplicities, assemblages, and rhizomatic forms.

35: Giedion, Siegfried. (1948). Mechanization Takes Command: A Contribution to Anonymous History. University of Minnesota Press, 419.

36: Giedion, Siegfried. (1948). Mechanization Takes Command: A Contribution to Anonymous History. University of Minnesota Press, 424.

37: Barthes, Roland. (1977). Image Music Text. Hill and Wang, 148.

38: Giedion, Siegfried. (1948). Mechanization Takes Command: A Contribution to Anonymous History. University of Minnesota Press, 714.

39: Of course, AI technology has been in development for decades and already found its way into many mainstream applications, but because of the rapid spread of text-to-image and language generation models, it feels new to many, maybe even most.

40: Marinetti, F. T. (1968). "The Foundation and Manifesto of Futurism," 1980 as reprinted in Herschel B. Chipp, Theories of Modern Art, UC Press, Berkeley.

41: Burke, Edmund. (1757). A Philosophical Inquiry into the Origin of Our Ideas of the Sublime and Beautiful.

42: Kant, Immanuel. (1764). Observations on the Feeling of the Beautiful and the Sublime.

43: For example: the data center of CERN (The European Center for Nuclear Research) in Geneva has more than 487 PT (Peta Byte, One Petabyte is 1024 Terabyte of data)

44: Schopenhauer, Arthur. The World as Will and Representation, § 39 "Fullest Feeling of Sublime – Immensity of Universe's extent or duration. (Pleasure from knowledge of observer's nothingness and oneness with Nature)."

45: Weng, Lilian. (2021). What are diffusion models? Lil'Log. https://lilianweng.github.io/posts/2021-07-11-diffusion-models/.

46: Pietruszka, M., Borchmann, L., and Garncarek, L. (2020). Sparsifying Transformer Models with Trainable Representation Pooling. ArXiv. /abs/2009.05169

47: Hogenboom, Katja. (2014). "The Possibilities of Emancipating Architecture – Strategies of Estrangement" Higher Seminar; Presentation and discussion of my PhD research in seminar with Inez Weisman, Umeå School of Architecture, Umeå, Sweden, 28 October, 2014.

48: Eisenman, P. (1976). Post-functionalism. Oppositions, 6: 19–21.

49: Snooks, Roland and Harper, Laura. (2020). Printed Assemblages: A Co-Evolution of Composite Tectonics and Additive Manufacturing Techniques. 10.2307/j.ctv13xpsvw.31.

50: Young, Michael. (2015). The Estranged Object: Young & Ayata, Graham Foundation for Advanced Studies in the Fine Arts, Graham Foundation.

51: Hogenboom, Katja. (2014). "The Possibilities of Emancipating Architecture – Strategies of Estrangement"

Higher Seminar; Presentation and discussion of my PhD research in seminar with Inez Weisman, Umeå School of Architecture, Umeå, Sweden, 28 October, 2014.

52: Katja, Hogenboom. (2014). "The Possibilities of Emancipating Architecture – Strategies of Estrangement" Higher Seminar; Presentation and discussion of my PhD research in seminar with Inez Weisman, Umeå School of Architecture, Umeå, Sweden, 28 October, 2014.

53: Carpo, Mario. (2017). The Second Digital Turn, Design Beyond Intelligence, MIT Press.

54: 'tool making is not simply the making of instruments by animal bodies but also the making of a body by the tool', Malafouris, Lambros, How does thinking relate to tool making? Sage Journal, v.29 I2

55: Cache, Bernard. (2011). Projectiles. Architectural Association.

56: The first building described by Vitruvius in the De Architectura is the Tower of Winds by Andronicus of Cyrrhus, a structure that features a combination of sundials, a water clock, and a wind vane. The building was aedificatio (a building, with an interior and exterior), gnomonica (with its solar clocks), and mecanica (with its rotating planetarium) – the three fields of architecture identified in the first book of De Architectura. One single building in this way reifies a whole theory of architecture. According to Bernard Cache, the structure was conceived as a proto-computational machine, a machine to measure and transmit information.

57: 'Partes ipsius architecturae sunt aedificatio, gnomonice, machinatio', Marcus Vitruvius Pollio, De Architectura, book I (3,1).

58: Portoghesi, Paolo. (2014). Tecnica Curiosa. Medusa.

59: Mario, Carpo. (2011). The Alphabet and the Algorithm. MIT Press.

60: Evans, Robin. (1997). Translation from Drawing to Building and Other Essays, MIT Press.

61: In ancient Greek symmetria "agreement in dimensions, duc proportion, arrangement," from symmetros "having a common measure, even, proportionate," from assimilated form of syn- "together" + metron "measure".

62: "At the end of book VI of De Architectura, Vitruvius affirms that the choice of material is the prerogative of the client, while the skuilful execution of the project depends on the craftsman: hence the architect's only chance of glory lies in the seductiveness of 'the proportion and symmetry which enter the design'". Bernard Cache, Projectile, AA, 2011

63: Andre Chastel introduced the notion of "mathematical humanism" in his book Centri del Rinascimento: Arte italiana 1460–1500 (Milan: Feltrinelli, 1965). Chastel identifies three strands of humanism and specifies that the mathematical one "finds its most important base in Urbino" (41), noting that "the case of Luca Pacioli is not isolated: on the contrary, it well represents the intellectual environment of the quattrocento, an environment in which theory and practice walk hand in hand without, however, adapting themselves to one another perfectly" (47, 49).

64: Pacioli, Luca, De Divina, Proportione, Aboca, Museum and San, Sepolcro Rackusin, Byrna. (1977). "The Architectural Theory Of Luca Pacioli: De Divina Proportione, Chapter 54." Bibliothèque d'Humanisme et Renaissance 39(3):479–502. http://www.jstor.org/stable/20675777.

65: Cache, Bernard. (2007). The Computation of Vitruvius. AA Lecture.

66: Before the invention of printing, which guaranteed reliable precision in reproducing drawings, Vitruvius and, in the 15th century, Leon Battista Alberti used writings to transmit architectural protocols. While Alberti's set of instructions could be easily followed by any draftsman or stonecutter, as they described a mechanical process of making things; Vitruvius's focus was more on the process of establishing proportions by means of parametric relations between parts.

67: In his Rule of the Five Orders (ca.1562–63), Giacomo Barozzi da Vignola abandoned any verbal instructions and indicated proportional measurements by printing numbers directly onto the drawing plates. Vignola illustrated the order by quantifying with numbers the proportional relationship between the different parts. His method was based on a single unit of measurement, called "module." Vignola was the first one to include integers and fractions in his modular notations. In turn, the module was divided into parts. Following up on Vignola's Rule, Andrea Palladio made large use of numbers for the description of the Doric order in the first book of his Four Books of Architecture (1570). Palladio, however, never abandoned the written description: he often indulged in duplicating the same information in drawings and texts, allowing both the numerati and the more traditional architects to equally understand and study his books.

68: Descartes, René. (1637). La Géométrie. Translated by Orion Publishing Group. (2004).

69: Cache, Bernard. (2005). Objectile: 6 (Consequence Book Series on Fresh Architecture): v. 6 Paperback – 1 Sept. 2005.

70: Carpo, Mario. (2017). The Second Digital Turn. MIT Press

71: Witt, Andrew. (2022). Formulations. MIT Press.

72: "The word 'black box' is used by cyberneticians whenever a piece of machinery or a set command is too complex. In its place, they draw a little box about which they need to know nothing but its input and output". Bruno Latour, Science in Action: How to Follow Scientists and Engineers through Society, Cambridge, MA, Harvard University Press, 1987, 2–3

73: Witt, Andrew. (2021). Formulations. MIT Press.

74: Carpo, Mario. (2017). The Second Digital Turn, MIT Press.

75: Negroponte, Nicholas. (1970). The Architecture Machine. MIT Press.

76: Bratton, Benjamin. (2015). The Stack: On Software and Sovereignty. MIT Press.

Vestures

1: Laclau, E. (1996b). Why Do Empty Signifiers Matter to Politics? In: E. Laclau (ed) Emancipation(s). London and New York: Verso, pp. 36–46.

2: Mostafavi, S. (2021). Hybrid Intelligence in Architectural Robotic Materialization (HI-ARM), A+ BE| Architecture and the Built Environment, no. 12, 2021, pp. 1–266. https://doi.org/10.7480/abe.2021.12.5799

3: Gary, M. (2023). The Dark Risk of Large Language Models in THE WIRED WORLD in 2023. Last date access 3/11/2023 https://www.wired.com/story/large-language-models-artificial-intelligence/

4: Agarwal. P. (2023). Emotional AI Is No Substitute for Empathy in The Wired World in 2023. Last date access 3/11/2023 https://www.wired.com/story/artificial-intelligence-empathy/

5: Eco, U. (1976). A Theory of Semiotics. Indiana: Indiana University Press.

6: Laclau, E. (2005). On Populist Reason. London: Verso.

7: Kozlowski, M., Mehan, A., and Nawratek, K. (2020). Kuala Lumpur: Community, infrastructure and urban inclusivity. Routledge.

8: Laclau, E. (1996a). Emancipation(s). London and New York: Verso.

9: Laclau, E., and Mouffe, C. (2005). Hegemony and Socialist Strategy: Towards a Radical Democratic Politics. London: Verso.

10: Matias del, Campo. (2022), Neural Architecture: Design and Artificial Intelligence First ed. Novato CA: Applied Research and Design Publishing an imprint of ORO Editions.

Estrangements

1: Battelli, C. (2022). Aladdin's lamp: Artificial intelligence, Architecture and Imagination, https://www.metalocus.es/en/news/aladdins-lamp-artificial-intelligence-architecture-and-imagination?fbclid=IwAR3dDX1aykiCFqNYlJ5Znr9Mt8egKv6NICy4iMzY_UPPA8soS1kR6525QzA

2: Vid. Corbin, H. (1958). L'Imagination créatrice dans le Soufisme d' Ibn Arabî, Flammarion, Zurich.

3: Garin, E. (1988). Phantasia e Imaginatio tra Marsilio Ficino e Piero Pompanazzi, in AA.VV. (a cura de Fattori M. Bianchi M.), PhantasiaImaginatio, Edizioni dell' Ateneo, Roma, pg.5

4: De Rosa, G. (1997). Il concetto di Immaginazione nel pensiero di Giordano Bruno, La città del Sole, Napoli.

5: Betsky, A. (2002). The Voyage Begins: Using Midjourney in Architecture, Aaron Betsky on artificial intelligence-driven software and the game-changing work of architect Cesare Battelli. https://www.architectmagazine.com/design/the-voyage-begins-using-midjourney-in-architecture_o?fbclid=IwAR0ly0WcyWppIIZDLIPZuysVz5_QGb1sfgUlyLVy7Cc7Ke9pZswNWmxOuI4

6: AA.VV. (1981). The Bachelor Machines, Rizzoli, New York, pg. 5

7: Kafka, F. (1919). In der Strafkolonie, Kurt Wolf Verlag, Lipsia.

8: Ishiguro, K. (2021). Klara and the Sun. London: Faber & Faber. pp. 117–118.

9: Corbusier, Le. (1920–1922). Drawing of Still Life [Drawing].

10: Corbusier, Le. (1929). Villa Savoye [Building]. Poissy, France.

11: Koh, Immanuel. (2021). 3D-GAN Housing [Exhibition]. Venice Architecture Biennale.

12: Klingemann, M. (2019). Memories of Passersby I [Artwork]. Sotheby's Auction.

13: McDonald, K. (2018). How to Recognize Fake AI-Generated Images.

14: Hertzmann, A. (2018). The AI Art Manifesto.

15: Guanzhong, Wu. (1992). Pandas [Ink on paper]. National Gallery Singapore Collection.

16: Picasso, P. (1950). Chouette Femme [Painting]. Private Collection.

17: Picasso, P. (1951). Owl [Painting]. Museum of Modern Art, New York.

18: Koh, Immanuel. (2021) '3D-GAN-Ar-Chair-tecture' [Exhibition]. Venice Architecture Biennale.

19: Koh, Immanuel. (2023). '3D-Diffusion-Ar-Chair-TEXTure' [Commission]. Asian Civilisation Museum, Singapore.

20: Ligon, G. (1993). The Runaways [Series of lithographs]. New York: Printed Matter, Inc.

21: Manferdini, Elena. (2023). Pick me, Atelier Manferdini.

22: Ngai, Sianne. (2022). Our Aesthetic Categories: Zany, Cute, Interesting. Cambridge, Massachusetts: Harvard University Press, p. 67.

23: Crawford, Kate and Paglen, Trevor. (2019). "Excavating AI: The Politics of Training Sets for Machine Learning (September 19, 2019), https://excavating.ai.

24: Rehms, Casey Michael. (2019). "SCI-ARC Vertical Studio 2019: American Politics as Design Methodology." SCI-Arc, Southern California Institute of Architecture.

25: Goodwin, Dario. (2017). "Spotlight: John Hejduk." ArchDaily, 6 Mar. 2017, www.archdaily.com/866691/spotlight-john-hejduk.

26: Goodwin, Dario. (2017). "Spotlight: John Hejduk." ArchDaily, 6 Mar. 2017, www.archdaily.com/866691/spotlight-john-hejduk.

27: The Associated Press. (2014). "Gonzo journalism." The Guardian, 9 August 2014, https://www.theguardian.com/media/greenslade/2014/aug/09/gonzo-journalism.

28: DeviantArt. (2023). About DeviantArt, https://www.deviantart.com/about. 2023.

29: Eliot, T. S. (1920). The Sacred Wood: Essays on Poetry and Criticism. Methuen & Co. Ltd.

30: Dick, P. K. (1969). The Galactic Pot Healer. Berkley Books.

31: Konami Digital Entertainment. (2009). LovePlus [Video game]. Nintendo DS.

32: Herzog, W. (Director). (2019). Family Romance LLC [Motion picture]. United States: MUBI.

33: Dick, P. K. (1996). Do Androids Dream of Electric Sheep? Paperback.

34: McLean, D. (1971). American Pie [Recorded by Don McLean]. New York: Record Plant Studios

35: Eliot, T. S. (1920). The Sacred Wood: Essays on Poetry and Criticism. London: Methuen & Co. Ltd.

Bibliography

Roland, Barthes and Stafford, Andy. (2006). The Language of Fashion English ed. Oxford: Berg.

Barthes, R. (1972). Mythologies. New York: Hill and Wang.

Barthes, R. (1968). Elements of Semiology. New York: Hill & Wang.

Baudrillard, J. (1994) Simulacra and Simulation. Ann Arbor, MI: The University of Michigan Press.

Beyaert-Geslin, A. (2012) Sémiotique du design. Paris: Presses Universitaires de France.

Benjamin, L. (1997). Introduction in Talking Heads: Language, Metalanguage, and the Semiotics of Subjectivity. New York, Duke University Press.

Berry, W. (2001). A Place on Earth. Counterpoint.

Braudel, F. (1986). Civilization and Capitalism 15th-18th Century. New York: Harper & Row.

Braverman, H. (1974). Labor and Monopoly Capital. the Degradation of Work in the Twentieth Century. New York: Monthly Review Press.

Bridle, J. (2018). New Dark Age : Technology and the End of the Future. London: Verso.

Carpo, M. (2017). The Second Digital Turn, Design Beyond Intelligence. Cambridge, MA/London: MIT Press

Casey, Edward S. (1997). The Fate of Place: A Philosophical History. 1st ed. University of California Press.

Corbin H. (1958). L’ímagination Créatrice Dans Le Soufisme D’ibn □arabi. 2 nd. Paris: Flammarion.

Crawford, K. (2021). The Atlas of Ai. Yale University Press.

De Rosa, G. (1997). Il concetto di Immaginazione nel pensiero di Giordano Bruno. Napoli: La città del Sole.

del Campo, M. (2022). Neural Architecture: Design and Artificial Intelligence. 1st ed. Novato CA: Applied Research and Design Publishing an imprint of ORO Editions.

Dick, P. K. (1969). The Galactic Pot Healer. Berkley Books.

Dick, P. K. (1996). Do Androids Dream of Electric Sheep? Paperback.

Laclau, E. ed. (1996). Emancipation(s). London and New York: Verso.

Eco, U. (1976). A Theory of Semiotics. Indiana: Indiana University Press.

Eliot, T. S. (1920). The Sacred Wood: Essays on Poetry and Criticism. Methuen & Co. Ltd.

Foucault, M. (1977). Discipline and Punish: The Birth of the Prison. New York : Pantheon Books.

Garin E. (1988). Phantasia e Imaginatio tra Marsilio Ficino e Piero Pompanazzi, in AA.VV. (a cura de Fattori, M. and Bianchi, M.), PhantasiaImaginatio, Edizioni dell’ Ateneo, Roma.

Gerald, M. L. (2006). The Turk Chess Automaton. McFarland & Co Inc Pub.

Ishiguro, K. (2021). Klara and the Sun. London: Faber & Faber.

Malpas, J. (2012). Heidegger and the Thinking of Place: Explorations in the Topology of Being. The MIT Press, 2012

Manning, C. D. and Schütze, H. (1999). Foundations of Statistical Natural Language Processing. Cambridge, MA The MIT Press.

Michels, K. (1989). Der Sinn der Unordnung. Arbeitsformen im Atelier Le Corbusier. Braunschweig/ Wiesbaden: Vieweg.

Kafka, F. (1919). In der Strafkolonie. Lipsia: Kurt Wolf Verlag

Koopman, C. (2019). How We Became Our Data: A Genealogy of the Informational Person. Chicago, IL University of Chicago Press.

Kozlowski, M., Mehan, A., and Nawratek, K. (2020). Kuala Lumpur: Community, infrastructure and urban inclusivity. Routledge.

Laclau, E. and Mouffe, C. (2005). Hegemony and Socialist Strategy: Towards a Radical Democratic Politics. London: Verso.

Laclau, E. (2005). On Populist Reason. London: Verso.

Le Bot, M. Brock, B. Carrouges, M. De Certeau, M. Clair, J. Gorsen, P. Le Macchine Celebi / The Bachelor Machines. Rizzoli: New York.

Michels, K. (1989). Der Sinn der Unordnung. Arbeitsformen im Atelier Le Corbusier Braunschweig/ Wiesbaden: Vieweg

Ngai, Sianne. (2012). Our Aesthetic Categories: Zany, Cute, Interesting. Cambridge, MA Harvard University Press.

Onuoha, M. and Nucera D. (2018). A People's Guide to AI, Detroit.

Reizen, P.Z. (2018). Happiness for Human. New York: Grand Central Publishing.

Zuboff, S. (1989). In the Age of the Smart Machine the Future of Work and Power. New York Basic Books.

Zuboff, S. (2019). The Age of Surveillance Capitalism: The Fight for a Human Future at the New Frontier of Power. 1st ed. New York: PublicAffairs.

Index

Image Credits

Cover

Matias del Campo

Prologue

1:©

Suppositions

Matias del Campo

1:©
2:©

Bart Lootsma

1:© "The streets no longer lead to fashion's future; today trends break out on the internet." (Dean Kissick, 2013)
2:© Susan Sontag in a cave-like Art Nouvea

Sandra Manninger

1:© Labeling Farm, according to Midjourney. Prompt: Photography of an MTurk Digital Sweatshop, people of diverse races Labeling Images ~v 4
2:© Photograph with flash of protesting digital workers, members of the diverse cultural background Computational Proletariat, crowdworkers ~v 4
3:© Allegory of Fei Fei Li at the moment she conceived the idea to ImageNet. According to Midjourney. Prompt: photography of Fei Fei Li inventing ImageNet ~v 4
4:© Ritter von Kempelen's Mechanical Turk, 1771, image: Joseph Racknitz, 1789
5:© Lithium mining in the Atacama Desert, Chile (according to Midjourney) Prompt: photography of the Lithium Mine in the Atacama Desert in Chile, cinematic illumination, National Geographic magazine photography ~v 4

6:© Common House Dataset. Web interface for crowd sourcing of plans

7:© Annotators protesting for their rights, according to Midjourney. Prompt: photograph with flash of Digital Computer white collar Union Workers protesting on the street, holding their laptops in the air, members of the Filipino Computational Proletariat, men and women, ~v 4

Joy Knoblauch

1:© Mimi Onuoha and Diana Nuncera use design to educate humans about what AI is and what is at stake in its use through publications such as A People's Guide to AI from 2018.

2:© Gwynne Michelle and Midjourney, "Renaissance Tracksuits," on Midjourney Official Facebook page February 25, 2023.

3:© This image was part of an exploration of hospital aesthetics using the terms hospital, bed, healing, affordable, window, and intravenous. Joy Knoblauch with Midjourney, January 2023

4:© Joy Knoblauch + Midjourney, Architecture Theory and Method, January 2023.

Matias del Campo

1:© The 24 High School designed by SPAN using an Attentional Generative Adversarial Network (AttnGAN), One of the first attempts to use text to image in the context of architecture design.

2:© Cladogram of Diffusion models and its applications. Based on the work of Yang, Ling, Zhilong Zhang, Shenda Hong, Runsheng Xu, Yue Zhao, Yingxia Shao, Wentao Zhang, Ming Hsuan Yang and Bin Cui.

3:© Diagram of the basic functionality of Diffusion models. Destructuring data by adding noise and then using denoising functions to extract data from the noise.

4:© "A chimp making tools" result of the prompt in Midjourney AI.

5:© Ludwig Wittgenstein and John Locke discussing language in a cafe in Vienna. According to Midjourney

6:© "A house made of crude concrete on top of a rock", this is the prompt used to generate this image. Midjourney is a powerful tool for creating concept images, but can it do more?

Ryan Vincent Manning

1:© Ryan Vincent_1

2:© Ryan Vincent_2

3:© Ryan Vincent_3

4:© Ryan Vincent_4

5:© Ryan Vincent_5

6:© Ryan Vincent_6

7:© Ryan Vincent_7

Commorancies

Kory Bieg

1:© Blue and Gold Housing Series, Kory Bieg, 2022. Designed with Midjourney.
2:© Kory Bieg _2
3:© Bio Series, Kory Bieg, 2022. Designed with Midjourney.
4:© Kory Bieg _4
5:© Jellyfish Housing, Kory Bieg, 2023. Designed with Midjourney using an image blend of the Blue and Gold Housing Series and Bio Series.
6:© Kory Bieg _6
7:© Nature Halls, Kory Bieg, 2023. Designed with Midjourney
8:© Cluster Housing, Kory Bieg, 2023. Designed with Midjourney
9:© Nature Courts, Kory Bieg, 2023. Designed with Midjourney using an image blend of the Nature Halls Series and Cluster Housing Series.
10:© Nature Courts, Kory Bieg, 2023. Designed with Midjourney using an image blend of the Nature Halls Series and Cluster Housing Series.
11:© Nature Courts, Kory Bieg, 2023. Designed with Midjourney using an image blend of the Nature Halls Series and Cluster Housing Series.
12:© Nature Courts, Kory Bieg, 2023 Designed with Midjourney using an image blend of the Nature Halls Series and Cluster Housing Series.
13:© Nature Courts, Kory Bieg, 2023. Designed with Midjourney using an image blend of the Nature Halls Series and Cluster Housing Series.
14:© Nature Courts, Kory Bieg, 2023. Designed with Midjourney using an image blend of the Nature Halls Series and Cluster Housing Series.
15:© Nature Courts, Kory Bieg, 2023 Designed with Midjourney using an image blend of the Nature Halls Series and Cluster Housing Series.
16:© Nature Courts, Kory Bieg, 2023. Designed with Midjourney using an image blend of the Nature Halls Series and Cluster Housing Series.
17:© Nature Courts, Kory Bieg, 2023. Designed with Midjourney using an image blend of the Nature Halls Series and Cluster Housing Series.
18:© Nature Courts, Kory Bieg, 2023. Designed with Midjourney using an image blend of the Nature Halls Series and Cluster Housing Series.
19:© Nature Courts, Kory Bieg, 2023.
20:© Designed with Midjourney using an blend of the Nature Halls Series and Cluster Housing Series.
21:© Nature Courts, Kory Bieg, 2023 Designed with Midjourney using an image blend of the Nature Halls Series and Cluster Housing Series.
22:© Kory Bieg _22
23:© Machinic Housing Domains, Kory Bieg, 2023. Designed with Midjourney.
24:© Machinic Domains, Kory Bieg, Clay Odom, Benjamin Rice, 2018.
25:© Kory Bieg _25
26:© Kory Bieg _26
27:© Kory Bieg _27
28:© Plume. Design by Kory Bieg and Clay Odom. Austin, Texas, 2023.
29:© Lumifoil. Design by Kory Bieg and Clay Odom. Miami, Florida, 2015.
30:© Kory Bieg _30
31:© Kory Bieg _31
32:© Kory Bieg _32
33:© Berkshire House. Design by Kory Bieg. Alford, Massachusetts 2018.
34:© Kory Bieg _34

35:© Kory Bieg _35
36:© Kory Bieg _36
37:© Word Series. 3d model of image generated in Midjourney using letters to guide and predict form.
38:© Berkshire Cluster Buildings, Kory Bieg, 2023. Designed with Midjourney.
39:© Kory Bieg _39
40:© Kory Bieg _40
41:© Kory Bieg _41
42:© Kory Bieg _42
43:© Kory Bieg _43
44:© Kory Bieg _44
45:© Kory Bieg _45
46:© Black and White Peaks Series, Kory Bieg, 2022. Designed with Midjourney.
47:© Timber Peaks, Kory Bieg, 2023. Designed with Midjourney using a three image blend, including the Black and White Peaks Series, the Camouflage Series, and the Water Series.
48:© Camouflage Series, Kory Bieg, 2022. Designed with Midjourney.
49:© Water Series, Kory Bieg, 2022 Designed with Midjourney.
50:© Kory Bieg _50
51:© Housing Cluster Series, Kory Bieg, 2023. Designed with Midjourney.
52:© Housing Cluster Series, Kory Bieg, 2023. Designed with Midjourney.
53:© Housing Cluster Series, Kory Bieg, 2023 Designed with Midjourney.
54:© Housing Cluster Series, Kory Bieg, 2023 Designed with Midjourney.
55:© Housing Cluster Series, Kory Bieg, 2023.Designed with Midjourney.
56:© Housing Cluster Series, Kory Bieg, 2023. Designed with Midjourney.
57:© Housing Cluster Series, Kory Bieg, 2023. Designed with Midjourney.
58:© Housing Cluster Series, Kory Bieg, 2023. Designed with Midjourney.
59:© Housing Cluster Series, Kory Bieg, 2023. Designed with Midjourney.
60:© Housing Cluster Series, Kory Bieg, 2023. Designed with Midjourney.
61:© Housing Cluster Series, Kory Bieg, 2023. Designed with Midjourney.
62:© Housing Cluster Series, Kory Bieg, 2023. Designed with Midjourney.
63:© Housing Cluster Series, Kory Bieg, 2023. Designed with Midjourney.
64:© Housing Cluster Series, Kory Bieg, 2023. Designed with Midjourney.
65:© Housing Cluster Series, Kory Bieg, 2023. Designed with Midjourney.
66:© Housing Cluster Series, Kory Bieg, 2023.Designed with Midjourney.
67:© Housing Cluster Series, Kory Bieg, 2023. Designed with Midjourney.
68:© Housing Cluster Series, Kory Bieg, 2023. Designed with Midjourney.
69:© Housing Cluster Series, Kory Bieg, 2023. Designed with Midjourney.
70:© Housing Cluster Series, Kory Bieg, 2023. Designed with Midjourney.
71:© Housing Cluster Series, Kory Bieg, 2023. Designed with Midjourney.
72:© Housing Cluster Series, Kory Bieg, 2023. Designed with Midjourney.
73:© Housing Cluster Series, Kory Bieg, 2023. Designed with Midjourney.
74:© Housing Cluster Series, Kory Bieg, 2023. Designed with Midjourney.

75:© Housing Cluster Series, Kory Bieg, 2023. Designed with Midjourney.
76:© Housing Cluster Series, Kory Bieg, 2023. Designed with Midjourney.
77:© Wood Facade Series, Kory Bieg, 2022. Designed with Midjourney.
78:© Bio Series, Kory Bieg, 2022. Designed with Midjourney.
79:© Bio Wood Series, Kory Bieg, 2023. Designed with Midjourney using a three image blend, including the Wood Facade Series, the Bio Series, and a photograph of Parafish.
80:© Bio Wood Series, Kory Bieg, 2023. Designed with Midjourney using a three image blend, including the Wood Facade Series, the Bio Series, and a photograph of Parafish.
81:© Bio Wood Series, Kory Bieg, 2023. Designed with Midjourney using a three image blend, including the Wood Facade Series, the Bio Series, and a photograph of Parafish.
82:© Bio Wood Series, Kory Bieg, 2023. Designed with Midjourney using a three image blend, including the Wood Facade Series, the Bio Series, and a photograph of Parafish.
83:© Bio Wood Series, Kory Bieg, 2023. Designed with Midjourney using a three image blend, including the Wood Facade Series, the Bio Series, and a photograph of Parafish.
84:© Bio Wood Series, Kory Bieg, 2023. Designed with Midjourney using a three image blend, including the Wood Facade Series, the Bio Series, and a photograph of Parafish.
85:© Bio Wood Series, Kory Bieg, 2023. Designed with Midjourney using a three image blend, including the Wood Facade Series, the Bio Series, and a photograph of Parafish.
86:© Bio Wood Series, Kory Bieg, 2023. Designed with Midjourney using a three image blend, including the Wood Facade Series, the Bio Series, and a photograph of Parafish.
87:© Kory Bieg _87
88:© Kory Bieg _88
89:© Kory Bieg _89
90:© Kory Bieg _90
91:© Kory Bieg _91
92:© Kory Bieg _92
93:© Kory Bieg _93
94:© Kory Bieg _94
95:© Kory Bieg _95
96:© Kory Bieg _96
97:© Kory Bieg _97
98:© Kory Bieg _98
99:© Kory Bieg _99
100:© Kory Bieg _100
101:© Kory Bieg _101
102:© Kory Bieg _102
103:© Kory Bieg _103
104:© Kory Bieg _104
105:© Kory Bieg _105
106:© Kory Bieg _106
107:© Kory Bieg _107
108:© Kory Bieg _108
109:© Kory Bieg _109
110:© Kory Bieg _110
111:© Kory Bieg _111
112:© Kory Bieg _112

Soomen Hahm & Hanjun Kim

1:© Steampunk Pavilion Tallinn, Estonia 2019 Designed by Gwyllim Jahn & Camero Newnham (Fologram), SoomeenHahm Design, Igor Pantic

2:© Soomeen Hahm_2
3:© Soomeen Hahm_3
4:© Soomeen Hahm_4
5:© Soomeen Hahm_5
6:© Soomeen Hahm_6
7:© Soomeen Hahm_7
8:© Soomeen Hahm_8
9:© Soomeen Hahm_9
10:© Soomeen Hahm_10
11:© Soomeen Hahm_11
12:© Soomeen Hahm_12
13:© Soomeen Hahm_13
14:© Soomeen Hahm_14
15:© Soomeen Hahm_15
16:© Soomeen Hahm_16
17:© Soomeen Hahm_17
18:© Soomeen Hahm_18
19:© 2021: A Steam Odyssey SCI-Arc Exhibition, US, 2021 Project Credit: SoomeenHahm Design & Igor Pantic
20:© Soomeen Hahm_20
21:© Soomeen Hahm_21
22:© Soomeen Hahm_22
23:© Soomeen Hahm_23
24:© Soomeen Hahm_24
25:© Soomeen Hahm_25
26:© Soomeen Hahm_26
27:© Soomeen Hahm_27
28:© Soomeen Hahm_28
29:© Soomeen Hahm_29
30:© Soomeen Hahm_30
31:© Soomeen Hahm_31
32:© SpirallingTangle 2022 by Won Jae Lee, Sizhe Lu, Yilong Chen, Yangmin Su Instructor: Soomeen Hahm
33:© Soomeen Hahm_33
34:© Soomeen Hahm_34
35:© Soomeen Hahm_35
36:© Soomeen Hahm_36
37:© Soomeen Hahm_37
38:© Soomeen Hahm_38
39:© Soomeen Hahm_39
40:© Soomeen Hahm_40
41:© Soomeen Hahm_41
42:© Soomeen Hahm_42
43:© Soomeen Hahm_43
44:© Soomeen Hahm_44
45:© Soomeen Hahm_45
46:© Soomeen Hahm_46
47:© Soomeen Hahm_47
48:© Soomeen Hahm_48
49:© Soomeen Hahm_49
50:© Soomeen Hahm_50
51:© Soomeen Hahm_51
52:© Soomeen Hahm_52
53:© Soomeen Hahm_53
54:© Soomeen Hahm_54
55:© Soomeen Hahm_55
56:© Soomeen Hahm_56
57:© Soomeen Hahm_57
58:© Soomeen Hahm_58
59:© Soomeen Hahm_59
60:© Soomeen Hahm_60
61:© Soomeen Hahm_61
62:© Soomeen Hahm_62
63:© Soomeen Hahm_63
64:© Soomeen Hahm_64
65:© Soomeen Hahm_65
66:© SteelPunk 2022 by Alejandro Aguilera, Abhishek Kadian, James Chidiac, Jack Freedman Instructor: Soomeen Hahm

67:© Soomeen Hahm_67
68:© Soomeen Hahm_68
69:© Soomeen Hahm_69
70:© Soomeen Hahm_70
71:© Soomeen Hahm_71
72:© Soomeen Hahm_72
73:© Soomeen Hahm_73
74:© Soomeen Hahm_74
75:© Soomeen Hahm_75
76:© Soomeen Hahm_76
77:© Soomeen Hahm_77
78:© Soomeen Hahm_78
79:© Soomeen Hahm_79
80:© Soomeen Hahm_80
81:© Soomeen Hahm_81
82:© Soomeen Hahm_82
83:© Soomeen Hahm_83
84:© Soomeen Hahm_84
85:© Soomeen Hahm_85
86:© Soomeen Hahm_86
87:© Soomeen Hahm_87
88:© Soomeen Hahm_88
89:© Soomeen Hahm_89
90:© Soomeen Hahm_90
91:© Soomeen Hahm_91
92:© Soomeen Hahm_92
93:© Soomeen Hahm_93
94:© Soomeen Hahm_94
95:© Soomeen Hahm_95
96:© Soomeen Hahm_96
97:© Soomeen Hahm_97
98:© Soomeen Hahm_98
99:© Soomeen Hahm_99
100:© Soomeen Hahm_100
101:© Soomeen Hahm_101
102:© Soomeen Hahm_102
103:© Soomeen Hahm_103
104:© Soomeen Hahm_104
105:© Soomeen Hahm_105
106:© Soomeen Hahm_106
107:© Soomeen Hahm_107
108:© Soomeen Hahm_108
109:© Soomeen Hahm_109
110:© Soomeen Hahm_110
111:© Soomeen Hahm_111
112:© Soomeen Hahm_112
113:© Soomeen Hahm_113
114:© Augmented Ground Quebec, Canada 2020 Designed by Soomeen Hahm, Yumi Lee, JaeHeon Jung
115:© Soomeen Hahm_115
116:© Soomeen Hahm_116
117:© Soomeen Hahm_117
118:© Soomeen Hahm_118
119:© Soomeen Hahm_119
120:© Soomeen Hahm_120
121:© Soomeen Hahm_121
122:© Soomeen Hahm_122
123:© Soomeen Hahm_123
124:© Soomeen Hahm_124
125:© Soomeen Hahm_125
126:© Soomeen Hahm_126
127:© Soomeen Hahm_127
128:© Soomeen Hahm_128
129:© Soomeen Hahm_129
130:© Soomeen Hahm_130
131:© Soomeen Hahm_131
132:© Soomeen Hahm_132
133:© Soomeen Hahm_133
134:© Soomeen Hahm_134
135:© Soomeen Hahm_135
136:© Soomeen Hahm_136

137:© Soomeen Hahm_137
138:© Soomeen Hahm_138
139:© Soomeen Hahm_139
140:© Soomeen Hahm_140
141:© Soomeen Hahm_141
142:© Soomeen Hahm_142
143:© Soomeen Hahm_143
144:© Soomeen Hahm_144
145:© Soomeen Hahm_145
146:© Soomeen Hahm_146
147:© Soomeen Hahm_147
148:© Soomeen Hahm_148
149:© Soomeen Hahm_149
150:© Soomeen Hahm_150
151:© Soomeen Hahm_151
152:© Soomeen Hahm_152
153:© Soomeen Hahm_153
154:© Soomeen Hahm_154
155:© Soomeen Hahm_155
156:© Soomeen Hahm_156
157:© Soomeen Hahm_157
158:© Soomeen Hahm_158
159:© Soomeen Hahm_159
160:© Soomeen Hahm_160
161:© Soomeen Hahm_161
162:© Soomeen Hahm_162
163:© Soomeen Hahm_163
164:© Soomeen Hahm_164
165:© Soomeen Hahm_165
166:© Soomeen Hahm_166
167:© Soomeen Hahm_167
168:© Soomeen Hahm_168
169:© Soomeen Hahm_169
170:© Soomeen Hahm_170

Sandra Manninger

1:© The exhibition hall of the MAK located at the Weißkirchnerstraße.
2:© Section studies for The Doghouse (SPAN 2023)
3:© kunst-dokumentation.com/MAK
4:© The Doghouse: Axonometric view of the assembled model. The model consist of 110 CNC milled components. Drawing by Sang Wong Kang.
5:© The Doghouse: Assembly sequence for the setup of the model in the Museum of Applied Arts (MAK) in Vienna, Austria. Drawing by Sang Wong Kang 2023
6:© The Doghouse: Fabrication CNC cut components for assembly in the MAK (File created by Sang Wong Kang) SPAN 2023
7:© kunst-dokumentation.com/MAK
8:© kunst-dokumentation.com/MAK
9:© Mosaic - a gradient made of 20.264 individual images

Alicia Nahmad Vazquez

1:© Alicia Nahmad_1
2:© Alicia Nahmad_2
3:© Alicia Nahmad_3
4:© Alicia Nahmad_4
5:© Alicia Nahmad_5
6:© Alicia Nahmad_6
7:© Alicia Nahmad_7
8:© Alicia Nahmad_8
9:© Alicia Nahmad_9
10:© Alicia Nahmad_10
11:© Alicia Nahmad_11
12:© Alicia Nahmad_12

13:© Alicia Nahmad_13
14:© Alicia Nahmad_14
15:© Alicia Nahmad_15
16:© Alicia Nahmad_16
17:© Alicia Nahmad_17
18:© Alicia Nahmad_18
19:© Alicia Nahmad_19
20:© Alicia Nahmad_20
21:© Alicia Nahmad_21
22:© Alicia Nahmad_22
23:© Alicia Nahmad_23
24:© Alicia Nahmad_24
25:© Alicia Nahmad_25
26:© Alicia Nahmad_26
27:© Alicia Nahmad_27
28:© Alicia Nahmad_28
29:© Alicia Nahmad_29
30:© Alicia Nahmad_30
31:© Alicia Nahmad_31
32:© Alicia Nahmad_32
33:© Alicia Nahmad_33
34:© Alicia Nahmad_34
35:© Alicia Nahmad_35
36:© Alicia Nahmad_36
37:© Alicia Nahmad_37
38:© Alicia Nahmad_38
39:© Alicia Nahmad_39
40:© Alicia Nahmad_40
41:© Alicia Nahmad_41
42:© Alicia Nahmad_42
43:© Alicia Nahmad_43
44:© Alicia Nahmad_44
45:© Alicia Nahmad_45
46:© Alicia Nahmad_46
47:© Alicia Nahmad_47
48:© Alicia Nahmad_48
49:© Alicia Nahmad_49
50:© Alicia Nahmad_50
51:© Alicia Nahmad_51
52:© Alicia Nahmad_52
53:© Alicia Nahmad_53
54:© Alicia Nahmad_54
55:© Alicia Nahmad_55
56:© Alicia Nahmad_56
57:© Alicia Nahmad_57
58:© Alicia Nahmad_58
59:© Alicia Nahmad_59
60:© Alicia Nahmad_60
61:© Alicia Nahmad_61
62:© Alicia Nahmad_62
63:© Alicia Nahmad_63
64:© Alicia Nahmad_64
65:© Alicia Nahmad_65
66:© Alicia Nahmad_66
67:© Alicia Nahmad_67
68:© Alicia Nahmad_68
69:© Alicia Nahmad_69
70:© Alicia Nahmad_70
71:© Alicia Nahmad_71
72:© Alicia Nahmad_72
73:© Alicia Nahmad_73
74:© Alicia Nahmad_74
75:© Alicia Nahmad_75
76:© Alicia Nahmad_76
77:© Alicia Nahmad_77

Marco Vanucci

1:© Marco Vanucci_1
2:© Marco Vanucci_2
3:© Marco Vanucci_3
4:© Marco Vanucci_4

5:© Marco Vanucci_5
6:© Marco Vanucci_6
7:© Marco Vanucci_7
8:© Marco Vanucci_8
9:© Marco Vanucci_9
10:© Marco Vanucci_10
11:© Marco Vanucci_11
12:© Marco Vanucci_12
13:© Marco Vanucci_13
14:© Marco Vanucci_14
15:© Marco Vanucci_15
16:© Marco Vanucci_16
17:© Marco Vanucci_17
18:© Marco Vanucci_18
19:© Marco Vanucci_19
20:© Marco Vanucci_20
21:© Marco Vanucci_21
22:© Marco Vanucci_22
23:© Marco Vanucci_23
24:© Marco Vanucci_24
25:© Marco Vanucci_25
26:© Marco Vanucci_26
27:© Marco Vanucci_27
28:© Marco Vanucci_28
29:© Marco Vanucci_29
30:© Marco Vanucci_30
31:© Marco Vanucci_31
32:© Marco Vanucci_32
33:© Marco Vanucci_33
34:© Marco Vanucci_34
35:© Marco Vanucci_35
36:© Marco Vanucci_36
37:© Marco Vanucci_37
38:© Marco Vanucci_38
39:© Marco Vanucci_39
40:© Marco Vanucci_40
41:© Marco Vanucci_41
42:© Marco Vanucci_42
43:© Marco Vanucci_43
44:© Marco Vanucci_44
45:© Marco Vanucci_45
46:© Marco Vanucci_46
47:© Marco Vanucci_47
48:© Marco Vanucci_48
49:© Marco Vanucci_49
50:© Marco Vanucci_50
51:© Marco Vanucci_51
52:© Marco Vanucci_52
53:© Marco Vanucci_53
54:© Marco Vanucci_54
55:© Marco Vanucci_55
56:© Marco Vanucci_56
57:© Marco Vanucci_57
58:© Marco Vanucci_58
59:© Marco Vanucci_59
60:© Marco Vanucci_60
61:© Marco Vanucci_61

Vestures

Daniel Bolojan

1:© Daniel Bolojan_1
2:© Daniel Bolojan_2
3:© Daniel Bolojan_3
4:© Daniel Bolojan_4
5:© Daniel Bolojan_5
6:© Daniel Bolojan_6
7:© Daniel Bolojan_7
8:© Daniel Bolojan_8
9:© Daniel Bolojan_9

10:© Daniel Bolojan_10
11:© Daniel Bolojan_11
12:© Daniel Bolojan_12
13:© Daniel Bolojan_13
14:© Daniel Bolojan_14
15:© Daniel Bolojan_15
16:© Daniel Bolojan_16
17:© Daniel Bolojan_17
18:© Daniel Bolojan_18
19:© Daniel Bolojan_19
20:© Daniel Bolojan_20
21:© Daniel Bolojan_21
22:© Daniel Bolojan_22
23:© Daniel Bolojan_23
24:© Daniel Bolojan_24
25:© Daniel Bolojan_25
26:© Daniel Bolojan_26
27:© Daniel Bolojan_27
28:© Daniel Bolojan_28
29:© Daniel Bolojan_29
30:© Daniel Bolojan_30
31:© Daniel Bolojan_31
32:© Daniel Bolojan_32
33:© Daniel Bolojan_33
34:© Daniel Bolojan_34
35:© Daniel Bolojan_35
36:© Daniel Bolojan_36
37:© Daniel Bolojan_37
38:© Daniel Bolojan_38
39:© Daniel Bolojan_39
40:© Daniel Bolojan_40
41:© Daniel Bolojan_41
42:© Daniel Bolojan_42
43:© Daniel Bolojan_43
44:© Daniel Bolojan_44
45:© Daniel Bolojan_45
46:© Daniel Bolojan_46
47:© Daniel Bolojan_47
48:© Daniel Bolojan_48
49:© Daniel Bolojan_49
50:© Daniel Bolojan_50
51:© Daniel Bolojan_51
52:© Daniel Bolojan_52
53:© Daniel Bolojan_53
54:© Daniel Bolojan_54
55:© Daniel Bolojan_55
56:© Daniel Bolojan_56
57:© Daniel Bolojan_57
58:© Daniel Bolojan_58
59:© Daniel Bolojan_59
60:© Daniel Bolojan_60
61:© Daniel Bolojan_61
62:© Daniel Bolojan_62
63:© Daniel Bolojan_63
64:© Daniel Bolojan_64
65:© Daniel Bolojan_65
66:© Daniel Bolojan_66
67:© Daniel Bolojan_67
68:© Daniel Bolojan_68
69:© Daniel Bolojan_69
70:© Daniel Bolojan_70
71:© Daniel Bolojan_71
72:© Daniel Bolojan_72
73:© Daniel Bolojan_73
74:© Daniel Bolojan_74
75:© Daniel Bolojan_75
76:© Daniel Bolojan_76
77:© Daniel Bolojan_77
78:© Daniel Bolojan_78
79:© Daniel Bolojan_79
80:© Daniel Bolojan_80
81:© Daniel Bolojan_81

82:© Daniel Bolojan_82
83:© Daniel Bolojan_83
84:© Daniel Bolojan_84
85:© Daniel Bolojan_85
86:© Daniel Bolojan_86
87:© Daniel Bolojan_87
88:© Daniel Bolojan_88
89:© Daniel Bolojan_89
90:© Daniel Bolojan_90
91:© Daniel Bolojan_91
92:© Daniel Bolojan_92
93:© Daniel Bolojan_93
94:© Daniel Bolojan_94
95:© Daniel Bolojan_95
96:© Daniel Bolojan_96
97:© Daniel Bolojan_97
98:© Daniel Bolojan_98
99:© Daniel Bolojan_99
100:© Daniel Bolojan_100
101:© Daniel Bolojan_101
102:© Daniel Bolojan_102
103:© Daniel Bolojan_103
104:© Daniel Bolojan_104
105:© Daniel Bolojan_105
106:© Daniel Bolojan_106
107:© Daniel Bolojan_107
108:© Daniel Bolojan_108
109:© Daniel Bolojan_109
110:© Daniel Bolojan_110
111:© Daniel Bolojan_111
112:© Daniel Bolojan_112
113:© Daniel Bolojan_113
114:© Daniel Bolojan_114
115:© Daniel Bolojan_115
116:© Daniel Bolojan_116
117:© Daniel Bolojan_117
118:© Daniel Bolojan_118
119:© Daniel Bolojan_119
120:© Daniel Bolojan_120

Daniel Koehler

1:© Daniel Koehler_1
2:© Daniel Koehler_2
3:© Daniel Koehler_3
4:© Daniel Koehler_4
5:© Daniel Koehler_5
6:© Daniel Koehler_6
7:© Daniel Koehler_7
8:© Daniel Koehler_8
9:© Daniel Koehler_9
10:© Daniel Koehler_10
11:© Daniel Koehler_11
12:© Daniel Koehler_12
13:© Daniel Koehler_13
14:© Daniel Koehler_14
15:© Daniel Koehler_15
16:© Daniel Koehler_16
17:© Daniel Koehler_17
18:© Daniel Koehler_18
19:© Daniel Koehler_19
20:© Daniel Koehler_20
21:© Daniel Koehler_21
22:© Daniel Koehler_22
23:© Daniel Koehler_23
24:© Daniel Koehler_24
25:© Daniel Koehler_25
26:© Daniel Koehler_26
27:© Daniel Koehler_27
28:© Daniel Koehler_28
29:© Daniel Koehler_29

30:© Daniel Koehler_30
31:© Daniel Koehler_31
32:© Daniel Koehler_32
33:© Daniel Koehler_33
34:© Daniel Koehler_34
35:© Daniel Koehler_35
36:© Daniel Koehler_36
37:© Daniel Koehler_37
38:© Daniel Koehler_38
39:© Daniel Koehler_39
40:© Daniel Koehler_40
41:© Daniel Koehler_41
42:© Daniel Koehler_42
43:© Daniel Koehler_43
44:© Daniel Koehler_44
45:© Daniel Koehler_45
46:© Daniel Koehler_46
47:© Daniel Koehler_47
48:© Daniel Koehler_48

Andrew Kudless

1:© Urban Resolution, 2023. Made with Stable Diffusion 2.1 with the prompts "A Street in Los Angeles", "A Street in New York", "A Street in Cairo", "A Street in Delhi", and "A Street in Tokyo".
2:© Urban Resolution, 2023. Made with Stable Diffusion 2.1 with the prompts "A Street in Los Angeles", "A Street in New York", "A Street in Cairo", "A Street in Delhi", and "A Street in Tokyo".
3:© Urban Resolution, 2023. Made with Stable Diffusion 2.1 with the prompts "A Street in Los Angeles", "A Street in New York", "A Street in Cairo", "A Street in Delhi", and "A Street in Tokyo."
4:© Urban Resolution, 2023. Made with Stable Diffusion 2.1 with the prompts "A Street in Los Angeles", "A Street in New York", "A Street in Cairo", "A Street in Delhi", and "A Street in Tokyo."
5:© Urban Resolution, 2023. Made with Stable Diffusion 2.1 with the prompts "A Street in Los Angeles", "A Street in New York", "A Street in Cairo", "A Street in Delhi", and "A Street in Tokyo".
6:© Andrew Kudless_6
7:© Andrew Kudless_7
8:© Andrew Kudless_8
9:© Andrew Kudless_9
10:© Andrew Kudless_10
11:© Andrew Kudless_11
12:© Andrew Kudless_12
13:© Andrew Kudless_13
14:© Andrew Kudless_14
15:© Andrew Kudless_15
16:© Andrew Kudless_16
17:© Andrew Kudless_17
18:© Andrew Kudless_18
19:© Andrew Kudless_19
20:© Andrew Kudless_20
21:© Andrew Kudless_21
22:© Andrew Kudless_22
23:© Andrew Kudless_23
24:© Facade Study 3841804661, 2023. Made with Stable Diffusion 1.5 using image prompts from Midjourney v4
25:© Dripping Springs House, 2022. An imagined nonlinear panorama of the house's interior spaces created in Midjourney and blended together in Dall-E2.
26:© Dripping Springs House, 2022. An imagined nonlinear panorama of the house's interior spaces created in Midjourney and blended together in Dall-E 2.

27:© Dripping Springs House, 2022. An imagined nonlinear panorama of the house's interior spaces created in Midjourney and blended together in Dall-E 2.

28:© Andrew Kudless_28

Sina Mostafavi

1:© Sina Mostafavi _1

2:© Sina Mostafavi _2

3:© Sina Mostafavi and Asma Mehan De-Coding Visual Cliches and Verbal Biases

4:© Sina Mostafavi and Asma Mehan De-Coding Visual Cliches and Verbal Biases

5:© Sina Mostafavi and Asma Mehan De-Coding Visual Cliches and Verbal Biases

6:© Sina Mostafavi and Asma Mehan De-Coding Visual Cliches and Verbal Biases

7:© Sina Mostafavi and Asma Mehan De-Coding Visual Cliches and Verbal Biases

8:© Sina Mostafavi and Asma Mehan De-Coding Visual Cliches and Verbal Biases

9:© Sina Mostafavi and Asma Mehan De-Coding Visual Cliches and Verbal Biases

10:© Sina Mostafavi and Asma Mehan De-Coding Visual Cliches and Verbal Biases

11:© Sina Mostafavi and Asma Mehan De-Coding Visual Cliches and Verbal Biases

12:© Sina Mostafavi and Asma Mehan De-Coding Visual Cliches and Verbal Biases

13:© Sina Mostafavi and Asma Mehan De-Coding Visual Cliches and Verbal Biases

14:© Sina Mostafavi and Asma Mehan De-Coding Visual Cliches and Verbal Biases

15:© Sina Mostafavi and Asma Mehan De-Coding Visual Cliches and Verbal Biases

16:© Sina Mostafavi and Asma Mehan De-Coding Visual Cliches and Verbal Biases

17:© Sina Mostafavi and Asma Mehan De-Coding Visual Cliches and Verbal Biases

18:© Sina Mostafavi and Asma Mehan De-Coding Visual Cliches and Verbal Biases

19:© Sina Mostafavi and Asma Mehan De-Coding Visual Cliches and Verbal Biases

20:© Sina Mostafavi and Asma Mehan De-Coding Visual Cliches and Verbal Biases

21:© Sina Mostafavi and Asma Mehan De-Coding Visual Cliches and Verbal Biases

22:© Sina Mostafavi and Asma Mehan De-Coding Visual Cliches and Verbal Biases

23:© Sina Mostafavi and Asma Mehan De-Coding Visual Cliches and Verbal Biases

24:© Sina Mostafavi and Asma Mehan De-Coding Visual Cliches and Verbal Biases

25:© Sina Mostafavi and Asma Mehan De-Coding Visual Cliches and Verbal Biases

26:© Sina Mostafavi and Asma Mehan De-Coding Visual Cliches and Verbal Biases

27:© Sina Mostafavi and Asma Mehan De-Coding Visual Cliches and Verbal Biases

28:© Sina Mostafavi and Asma Mehan De-Coding Visual Cliches and Verbal Biases

29:© Sina Mostafavi and Asma Mehan De-Coding Visual Cliches and Verbal Biases

30:© Sina Mostafavi and Asma Mehan De-Coding Visual Cliches and Verbal Biases

31:© Sina Mostafavi and Asma Mehan De-Coding Visual Cliches and Verbal Biases

32:© Sina Mostafavi and Asma Mehan De-Coding Visual Cliches and Verbal Biases

33:© Sina Mostafavi and Asma Mehan De-Coding Visual Cliches and Verbal Biases

34:© Sina Mostafavi and Asma Mehan De-Coding Visual Cliches and Verbal Biases
35:© Sina Mostafavi and Asma Mehan De-Coding Visual Cliches and Verbal Biases
36:© Sina Mostafavi and Asma Mehan De-Coding Visual Cliches and Verbal Biases
37:© Sina Mostafavi and Asma Mehan De-Coding Visual Cliches and Verbal Biases
38:© Sina Mostafavi and Asma Mehan De-Coding Visual Cliches and Verbal Biases
39:© Sina Mostafavi and Asma Mehan De-Coding Visual Cliches and Verbal Biases
40:© Sina Mostafavi and Asma Mehan De-Coding Visual Cliches and Verbal Biases
41:© Sina Mostafavi and Asma Mehan De-Coding Visual Cliches and Verbal Biases
42:© Sina Mostafavi and Asma Mehan De-Coding Visual Cliches and Verbal Biases
43:© Sina Mostafavi and Asma Mehan De-Coding Visual Cliches and Verbal Biases
44:© Sina Mostafavi and Asma Mehan De-Coding Visual Cliches and Verbal Biases
45:© Sina Mostafavi and Asma Mehan De-Coding Visual Cliches and Verbal Biases
46:© Sina Mostafavi and Asma Mehan De-Coding Visual Cliches and Verbal Biases
47:© Sina Mostafavi and Asma Mehan De-Coding Visual Cliches and Verbal Biases
48:© Sina Mostafavi and Asma Mehan De-Coding Visual Cliches and Verbal Biases
49:© Sina Mostafavi and Asma Mehan De-Coding Visual Cliches and Verbal Biase
50:© Sina Mostafavi and Asma Mehan De-Coding Visual Cliches and Verbal Biases
51:© Sina Mostafavi and Asma Mehan De-Coding Visual Cliches and Verbal Biases
52:© Sina Mostafavi and Asma Mehan De-Coding Visual Cliches and Verbal Biases
53:© Sina Mostafavi and Asma Mehan De-Coding Visual Cliches and Verbal Biases
54:© Sina Mostafavi and Asma Mehan De-Coding Visual Cliches and Verbal Biases
55:© Sina Mostafavi and Asma Mehan De-Coding Visual Cliches and Verbal Biases
56:© Sina Mostafavi and Asma Mehan De-Coding Visual Cliches and Verbal Biases
57:© Sina Mostafavi and Asma Mehan De-Coding Visual Cliches and Verbal Biases
58:© Sina Mostafavi and Asma Mehan De-Coding Visual Cliches and Verbal Biases
59:© Sina Mostafavi and Asma Mehan De-Coding Visual Cliches and Verbal Biases
60:© Sina Mostafavi and Asma Mehan De-Coding Visual Cliches and Verbal Biases
61:© Sina Mostafavi and Asma Mehan De-Coding Visual Cliches and Verbal Biases
62:© Sina Mostafavi and Asma Mehan De-Coding Visual Cliches and Verbal Biases
63:© Sina Mostafavi and Asma Mehan De-Coding Visual Cliches and Verbal Biases
64:© Sina Mostafavi and Asma Mehan De-Coding Visual Cliches and Verbal Biases
65:© Sina Mostafavi and Asma Mehan De-Coding Visual Cliches and Verbal Biases
66:© Sina Mostafavi and Asma Mehan De-Coding Visual Cliches and Verbal Biases
67:© Sina Mostafavi and Asma Mehan De-Coding Visual Cliches and Verbal Biases
68:© Sina Mostafavi and Asma Mehan De-Coding Visual Cliches and Verbal Biases
69:© Sina Mostafavi and Asma Mehan De-Coding Visual Cliches and Verbal Biases

70:© Sina Mostafavi and Asma Mehan De-Coding Visual Cliches and Verbal Biases
71:© Sina Mostafavi and Asma Mehan De-Coding Visual Cliches and Verbal Biases
72:© Sina Mostafavi and Asma Mehan De-Coding Visual Cliches and Verbal Biases
73:© Sina Mostafavi and Asma Mehan De-Coding Visual Cliches and Verbal Biases
74:© Sina Mostafavi and Asma Mehan De-Coding Visual Cliches and Verbal Biases
75:© Sina Mostafavi and Asma Mehan De-Coding Visual Cliches and Verbal Biases
76:© Sina Mostafavi and Asma Mehan De-Coding Visual Cliches and Verbal Biases
77:© Sina Mostafavi and Asma Mehan De-Coding Visual Cliches and Verbal Biases
78:© Sina Mostafavi and Asma Mehan De-Coding Visual Cliches and Verbal Biases
79:© Sina Mostafavi and Asma Mehan De-Coding Visual Cliches and Verbal Biases
80:© Sina Mostafavi and Asma Mehan De-Coding Visual Cliches and Verbal Biases
81:© Sina Mostafavi and Asma Mehan De-Coding Visual Cliches and Verbal Biases
82:© Sina Mostafavi and Asma Mehan De-Coding Visual Cliches and Verbal Biases
83:© Sina Mostafavi and Asma Mehan De-Coding Visual Cliches and Verbal Biases
84:© Sina Mostafavi and Asma Mehan De-Coding Visual Cliches and Verbal Biases
85:© Sina Mostafavi and Asma Mehan De-Coding Visual Cliches and Verbal Biases
86:© Sina Mostafavi and Asma Mehan De-Coding Visual Cliches and Verbal Biases
87:© Sina Mostafavi and Asma Mehan De-Coding Visual Cliches and Verbal Biases
88:© Sina Mostafavi and Asma Mehan De-Coding Visual Cliches and Verbal Biases
89:© Sina Mostafavi and Asma Mehan De-Coding Visual Cliches and Verbal Biases
90:© Sina Mostafavi and Asma Mehan De-Coding Visual Cliches and Verbal Biases
91:© Sina Mostafavi and Asma Mehan De-Coding Visual Cliches and Verbal Biases
92:© Sina Mostafavi and Asma Mehan De-Coding Visual Cliches and Verbal Biases
93:© Sina Mostafavi and Asma Mehan De-Coding Visual Cliches and Verbal Biases
94:© Sina Mostafavi and Asma Mehan De-Coding Visual Cliches and Verbal Biases
95:© Sina Mostafavi and Asma Mehan De-Coding Visual Cliches and Verbal Biases
96:© Sina Mostafavi and Asma Mehan De-Coding Visual Cliches and Verbal Biases
97:© Sina Mostafavi and Asma Mehan De-Coding Visual Cliches and Verbal Biases
98:© Sina Mostafavi and Asma Mehan De-Coding Visual Cliches and Verbal Biases
99:© Sina Mostafavi and Asma Mehan De-Coding Visual Cliches and Verbal Biases
100:© Sina Mostafavi and Asma Mehan De-Coding Visual Cliches and Verbal Biases
101:© Sina Mostafavi and Asma Mehan De-Coding Visual Cliches and Verbal Biases
102:© Sina Mostafavi and Asma Mehan De-Coding Visual Cliches and Verbal Biases
103:© Sina Mostafavi and Asma Mehan De-Coding Visual Cliches and Verbal Biases
104:© Sina Mostafavi and Asma Mehan De-Coding Visual Cliches and Verbal Biases
105:© Sina Mostafavi and Asma Mehan De-Coding Visual Cliches and Verbal Biases

106:© Sina Mostafavi and Asma Mehan De-Coding Visual Cliches and Verbal Biases
107:© Sina Mostafavi and Asma Mehan De-Coding Visual Cliches and Verbal Biases
108:© Sina Mostafavi and Asma Mehan De-Coding Visual Cliches and Verbal Biases
109:© Sina Mostafavi and Asma Mehan De-Coding Visual Cliches and Verbal Biases
110:© Sina Mostafavi and Asma Mehan De-Coding Visual Cliches and Verbal Biases
111:© Sina Mostafavi and Asma Mehan De-Coding Visual Cliches and Verbal Biases
112:© Sina Mostafavi and Asma Mehan De-Coding Visual Cliches and Verbal Biases
113:© Sina Mostafavi and Asma Mehan De-Coding Visual Cliches and Verbal Biases
114:© Sina Mostafavi and Asma Mehan De-Coding Visual Cliches and Verbal Biases
115:© Sina Mostafavi and Asma Mehan De-Coding Visual Cliches and Verbal Biases
116:© Sina Mostafavi and Asma Mehan De-Coding Visual Cliches and Verbal Biases
117:© Sina Mostafavi and Asma Mehan De-Coding Visual Cliches and Verbal Biases
118:© Sina Mostafavi and Asma Mehan De-Coding Visual Cliches and Verbal Biases
119:© Sina Mostafavi and Asma Mehan De-Coding Visual Cliches and Verbal Biases
120:© Sina Mostafavi and Asma Mehan De-Coding Visual Cliches and Verbal Biases
121:© Sina Mostafavi and Asma Mehan De-Coding Visual Cliches and Verbal Biases
122:© Sina Mostafavi and Asma Mehan De-Coding Visual Cliches and Verbal Biases
123:© Sina Mostafavi and Asma Mehan De-Coding Visual Cliches and Verbal Biases
124:© Sina Mostafavi and Asma Mehan De-Coding Visual Cliches and Verbal Biases
125:© Sina Mostafavi and Asma Mehan De-Coding Visual Cliches and Verbal Biases
126:© Sina Mostafavi and Asma Mehan De-Coding Visual Cliches and Verbal Biases
127:© Sina Mostafavi and Asma Mehan De-Coding Visual Cliches and Verbal Biases
128:© Sina Mostafavi and Asma Mehan De-Coding Visual Cliches and Verbal Biases
129:© Sina Mostafavi and Asma Mehan De-Coding Visual Cliches and Verbal Biases
130:© Sina Mostafavi and Asma Mehan De-Coding Visual Cliches and Verbal Biases
131:© Sina Mostafavi and Asma Mehan De-Coding Visual Cliches and Verbal Biases
132:© Sina Mostafavi and Asma Mehan De-Coding Visual Cliches and Verbal Biases
133:© Sina Mostafavi and Asma Mehan De-Coding Cliches and Verbal Biases

Rasa Navasaityte

1:© Rasa Navasaityte_1
2:© Rasa Navasaityte_2
3:© Rasa Navasaityte_3
4:© Rasa Navasaityte_4
5:© Rasa Navasaityte_5
6:© Rasa Navasaityte_6
7:© Rasa Navasaityte_7
8:© Rasa Navasaityte_8
9:© Rasa Navasaityte_9
10:© Rasa Navasaityte_10
11:© Rasa Navasaityte_11
12:© Rasa Navasaityte_12

13:© Rasa Navasaityte_13
14:© Rasa Navasaityte_14
15:© Rasa Navasaityte_15
16:© Rasa Navasaityte_16
17:© Rasa Navasaityte_17
18:© Rasa Navasaityte_18
19:© Rasa Navasaityte_19
20:© Rasa Navasaityte_20
21:© Rasa Navasaityte_21
22:© Rasa Navasaityte_22
23:© Rasa Navasaityte_23
24:© Rasa Navasaityte_24
25:© Rasa Navasaityte_25
26:© Rasa Navasaityte_26
27:© Rasa Navasaityte_27
28:© Rasa Navasaityte_28
29:© Rasa Navasaityte_29
30:© Rasa Navasaityte_30
31:© Rasa Navasaityte_31
32:© Rasa Navasaityte_32
33:© Rasa Navasaityte_33
34:© Rasa Navasaityte_34
35:© Rasa Navasaityte_35
36:© Rasa Navasaityte_36
37:© Rasa Navasaityte_37
38:© Rasa Navasaityte_38
39:© Rasa Navasaityte_39
40:© Rasa Navasaityte_40
41:© Rasa Navasaityte_41
42:© Rasa Navasaityte_42
43:© Rasa Navasaityte_43
44:© Rasa Navasaityte_44
45:© Rasa Navasaityte_45
46:© Rasa Navasaityte_46
47:© Rasa Navasaityte_47
48:© Rasa Navasaityte_48
49:© Rasa Navasaityte_49
50:© Rasa Navasaityte_50
51:© Rasa Navasaityte_51
52:© Rasa Navasaityte_52
53:© Rasa Navasaityte_53
54:© Rasa Navasaityte_54
55:© Rasa Navasaityte_55
56:© Rasa Navasaityte_56
57:© Rasa Navasaityte_57
58:© Rasa Navasaityte_58
59:© Rasa Navasaityte_59
60:© Rasa Navasaityte_60
61:© Rasa Navasaityte_61
62:© Rasa Navasaityte_62
63:© Rasa Navasaityte_63
64:© Rasa Navasaityte_64
65:© Rasa Navasaityte_65
66:© Rasa Navasaityte_66
67:© Rasa Navasaityte_67
68:© Rasa Navasaityte_68
69:© Rasa Navasaityte_69
70:© Rasa Navasaityte_70
71:© Rasa Navasaityte_71
72:© Rasa Navasaityte_72
73:© Rasa Navasaityte_73
74:© Rasa Navasaityte_74
75:© Rasa Navasaityte_75
76:© Rasa Navasaityte_76
77:© Rasa Navasaityte_77
78:© Rasa Navasaityte_78
79:© Rasa Navasaityte_79
80:© Rasa Navasaityte_80
81:© Rasa Navasaityte_81
82:© Rasa Navasaityte_82
83:© Rasa Navasaityte_83
84:© Rasa Navasaityte_84

85:© Rasa Navasaityte_85
86:© Rasa Navasaityte_86
87:© Rasa Navasaityte_87
88:© Rasa Navasaityte_88
89:© Rasa Navasaityte_89
90:© Rasa Navasaityte_90
91:© Rasa Navasaityte_91
92:© Rasa Navasaityte_92
93:© Rasa Navasaityte_93
94:© Rasa Navasaityte_94
95:© Rasa Navasaityte_95
96:© Rasa Navasaityte_96
97:© Rasa Navasaityte_97
98:© Rasa Navasaityte_98
99:© Rasa Navasaityte_99
100:© Rasa Navasaityte_100
101:© Rasa Navasaityte_101
102:© Rasa Navasaityte_102
103:© Rasa Navasaityte_103
104:© Rasa Navasaityte_104
105:© Rasa Navasaityte_105
106:© Rasa Navasaityte_106
107:© Rasa Navasaityte_107
108:© Rasa Navasaityte_108
109:© Rasa Navasaityte_109
110:© Rasa Navasaityte_110
111:© Rasa Navasaityte_111
112:© Rasa Navasaityte_112
113:© Rasa Navasaityte_113
114:© Rasa Navasaityte_114
115:© Rasa Navasaityte_115
116:© Rasa Navasaityte_116
117:© Rasa Navasaityte_117
118:© Rasa Navasaityte_118
119:© Rasa Navasaityte_119
120:© Rasa Navasaityte_120
121:© Rasa Navasaityte_121
122:© Rasa Navasaityte_122
123:© Rasa Navasaityte_123
124:© Rasa Navasaityte_124
125:© Rasa Navasaityte_125
126:© Rasa Navasaityte_126
127:© Rasa Navasaityte_127
128:© Rasa Navasaityte_128
129:© Rasa Navasaityte_129
130:© Rasa Navasaityte_130
131:© Rasa Navasaityte_131
132:© Rasa Navasaityte_132
133:© Rasa Navasaityte_133
134:© Rasa Navasaityte_134
135:© Rasa Navasaityte_135
136:© Rasa Navasaityte_136
137:© Rasa Navasaityte_137
138:© Rasa Navasaityte_138
139:© Rasa Navasaityte_139
140:© Rasa Navasaityte_140
141:© Rasa Navasaityte_141
142:© Rasa Navasaityte_142
143:© Rasa Navasaityte_143
144:© Rasa Navasaityte_144
145:© Rasa Navasaityte_145
146:© Rasa Navasaityte_146
147:© Rasa Navasaityte_147
148:© Rasa Navasaityte_148
149:© Rasa Navasaityte_149
150:© Rasa Navasaityte_150
151:© Rasa Navasaityte_151
152:© Rasa Navasaityte_152
153:© Rasa Navasaityte_153
154:© Rasa Navasaityte_154
155:© Rasa Navasaityte_155
156:© Rasa Navasaityte_156

157:© Rasa Navasaityte_157
158:© Rasa Navasaityte_158
159:© Rasa Navasaityte_159
160:© Rasa Navasaityte_160
161:© Rasa Navasaityte_161
162:© Rasa Navasaityte_162
163:© Rasa Navasaityte_163
164:© Rasa Navasaityte_164
165:© Rasa Navasaityte_165
166:© Rasa Navasaityte_166
167:© Rasa Navasaityte_167
168:© Rasa Navasaityte_168
169:© Rasa Navasaityte_169
170:© Rasa Navasaityte_170
171:© Rasa Navasaityte_171
172:© Rasa Navasaityte_172
173:© Rasa Navasaityte_173
174:© Rasa Navasaityte_174
175:© Rasa Navasaityte_175
176:© Rasa Navasaityte_176
177:© Rasa Navasaityte_177
178:© Rasa Navasaityte_178
179:© Rasa Navasaityte_179
180:© Rasa Navasaityte_180
181:© Rasa Navasaityte_181
182:© Rasa Navasaityte_182
183:© Rasa Navasaityte_183
184:© Rasa Navasaityte_184
185:© Rasa Navasaityte_185
186:© Rasa Navasaityte_186
187:© Rasa Navasaityte_187
188:© Rasa Navasaityte_188
189:© Rasa Navasaityte_189
190:© Rasa Navasaityte_190
191:© Rasa Navasaityte_191
192:© Rasa Navasaityte_192
193:© Rasa Navasaityte_193
194:© Rasa Navasaityte_194
195:© Rasa Navasaityte_195
196:© Rasa Navasaityte_196
197:© Rasa Navasaityte_197
198:© Rasa Navasaityte_198
199:© Rasa Navasaityte_199
200:© Rasa Navasaityte_200
201:© Rasa Navasaityte_201
202:© Rasa Navasaityte_202
203:© Rasa Navasaityte_203
204:© Rasa Navasaityte_204
205:© Rasa Navasaityte_205
206:© Rasa Navasaityte_206
207:© Rasa Navasaityte_207
208:© Rasa Navasaityte_208
209:© Rasa Navasaityte_209
210:© Rasa Navasaityte_210
211:© Rasa Navasaityte_211
212:© Rasa Navasaityte_212
213:© Rasa Navasaityte_213
214:© Rasa Navasaityte_214
215:© Rasa Navasaityte_215
216:© Rasa Navasaityte_216
217:© Rasa Navasaityte_217
218:© Rasa Navasaityte_218
219:© Rasa Navasaityte_219
220:© Rasa Navasaityte_220
221:© Rasa Navasaityte_221

Igor Pantic

1:© Igor Pantic_1
2:© Igor Pantic_2
3:© Igor Pantic_3

4:© Igor Pantic_4
5:© Igor Pantic_5
6:© Igor Pantic_6
7:© Igor Pantic_7
8:© Igor Pantic_8
9:© Igor Pantic_9
10:© Igor Pantic_10
11:© Igor Pantic_11
12:© Igor Pantic_12
13:© Igor Pantic_13
14:© Igor Pantic_14
15:© Igor Pantic_15
16:© Igor Pantic_16
17:© Igor Pantic_17
18:© Igor Pantic_18
19:© Igor Pantic_19
20:© Igor Pantic_20
21:© Igor Pantic_21
22:© Igor Pantic_22
23:© Igor Pantic_23
24:© Igor Pantic_24
25:© Igor Pantic_25
26:© Igor Pantic_26
27:© Igor Pantic_27
28:© Igor Pantic_28
29:© Igor Pantic_29
30:© Igor Pantic_30
31:© Igor Pantic_31
32:© Igor Pantic_32
33:© Igor Pantic_33
34:© Igor Pantic_34
35:© Igor Pantic_35
36:© Igor Pantic_36
37:© Igor Pantic_37
38:© Igor Pantic_38
39:© Igor Pantic_39
40:© Igor Pantic_40
41:© Igor Pantic_41
42:© Igor Pantic_42
43:© Igor Pantic_43
44:© Igor Pantic_44
45:© Igor Pantic_45
46:© Igor Pantic_46
47:© Igor Pantic_47
48:© Igor Pantic_48
49:© Igor Pantic_49
50:© Igor Pantic_50
51:© Igor Pantic_51
52:© Igor Pantic_52
53:© Igor Pantic_53
54:© Igor Pantic_54
55:© Igor Pantic_55
56:© Igor Pantic_56
57:© Igor Pantic_57
58:© Igor Pantic_58
59:© Igor Pantic_59
60:© Igor Pantic_60
61:© Igor Pantic_61
62:© Igor Pantic_62
63:© Igor Pantic_63
64:© Igor Pantic_64
65:© Igor Pantic_65
66:© Igor Pantic_66
67:© Igor Pantic_67
68:© Igor Pantic_68
69:© Igor Pantic_69
70:© Igor Pantic_70
71:© Igor Pantic_71
72:© Igor Pantic_72
73:© Igor Pantic_73
74:© Igor Pantic_74
75:© Igor Pantic_75

76:© Igor Pantic_76
77:© Igor Pantic_77
78:© Igor Pantic_78
79:© Igor Pantic_79
80:© Igor Pantic_80
81:© Igor Pantic_81
82:© Igor Pantic_82
83:© Igor Pantic_83
84:© Igor Pantic_84
85:© Igor Pantic_85
86:© Igor Pantic_86
87:© Igor Pantic_87
88:© Igor Pantic_88
89:© Igor Pantic_89
90:© Igor Pantic_90
91:© Igor Pantic_91
92:© Igor Pantic_92
93:© Igor Pantic_93
94:© Igor Pantic_94
95:© Igor Pantic_95
96:© Igor Pantic_96
97:© Igor Pantic_97
98:© Igor Pantic_98
99:© Igor Pantic_99
100:© Igor Pantic_100
101:© Igor Pantic_101
102:© Igor Pantic_102

Dustin White

1:© Dustin White_1
2:© Dustin White_2
3:© Dustin White_3
4:© Dustin White_4
5:© Dustin White_5
6:© Dustin White_6
7:© Dustin White_7
8:© Dustin White_8
9:© Dustin White_9
10:© Dustin White_10
11:© Dustin White_11
12:© Dustin White_12
13:© Dustin White_13
14:© Dustin White_14
15:© Dustin White_15
16:© Dustin White_16
17:© Dustin White_17
18:© Dustin White_18
19:© Dustin White_19
20:© Dustin White_20
21:© Dustin White_21
22:© Dustin White_22
23:© Dustin White_23
24:© Dustin White_24
25:© Dustin White_25
26:© Dustin White_26
27:© Dustin White_27
28:© Dustin White_28
29:© Dustin White_29
30:© Dustin White_30
31:© Dustin White_31
32:© Dustin White_32
33:© Dustin White_33
34:© Dustin White_34
35:© Dustin White_35
36:© Dustin White_36
37:© Dustin White_37
38:© Dustin White_38
39:© Dustin White_39
40:© Dustin White_40
41:© Dustin White_41
42:© Dustin White_42

43:© Dustin White_43
44:© Dustin White_44
45:© Dustin White_45
46:© Dustin White_46
47:© Dustin White_47
48:© Dustin White_48
49:© Dustin White_49
50:© Dustin White_50
51:© Dustin White_51
52:© Dustin White_52
53:© Dustin White_53
54:© Dustin White_54
55:© Dustin White_55
56:© Dustin White_56
57:© Dustin White_57
58:© Dustin White_58
59:© Dustin White_59
60:© Dustin White_60
61:© Dustin White_61
62:© Dustin White_62
63:© Dustin White_63
64:© Dustin White_64
65:© Dustin White_65
66:© Dustin White_66
67:© Dustin White_67
68:© Dustin White_68
69:© Dustin White_69
70:© Dustin White_70
71:© Dustin White_71
72:© Dustin White_72
73:© Dustin White_73
74:© Dustin White_74
75:© Dustin White_75
76:© Dustin White_76
77:© Dustin White_77
78:© Dustin White_78
79:© Dustin White_79
80:© Dustin White_80
81:© Dustin White_81
82:© Dustin White_82
83:© Dustin White_83
84:© Dustin White_84
85:© Dustin White_85
86:© Dustin White_86
87:© Dustin White_87
88:© Dustin White_88
89:© Dustin White_89
90:© Dustin White_90
91:© Dustin White_91
92:© Dustin White_92
93:© Dustin White_93
94:© Dustin White_94
95:© Dustin White_95
96:© Dustin White_96
97:© Dustin White_97
98:© Dustin White_98
99:© Dustin White_99
100:© Dustin White_100
101:© Dustin White_101
102:© Dustin White_102
103:© Dustin White_103
104:© Dustin White_104
105:© Dustin White_105
106:© Dustin White_106
107:© Dustin White_107
108:© Dustin White_108
109:© Dustin White_109
110:© Dustin White_110
111:© Dustin White_111
112:© Dustin White_112
113:© Dustin White_113
114:© Dustin White_114

115:© Dustin White_115
116:© Dustin White_116
117:© Dustin White_117
118:© Dustin White_118
119:© Dustin White_119
120:© Dustin White_120
121:© Dustin White_121
122:© Dustin White_122
123:© Dustin White_123
124:© Dustin White_124
125:© Dustin White_125
126:© Dustin White_126
127:© Dustin White_127
128:© Dustin White_128
129:© Dustin White_129
130:© Dustin White_130
131:© Dustin White_131
132:© Dustin White_132
133:© Dustin White_133
134:© Dustin White_134
135:© Dustin White_135
136:© Dustin White_136
137:© Dustin White_137
138:© Dustin White_138
139:© Dustin White_139
140:© Dustin White_140
141:© Dustin White_141
142:© Dustin White_142
143:© Dustin White_143
144:© Dustin White_144
145:© Dustin White_145
146:© Dustin White_146
147:© Dustin White_147
148:© Dustin White_148
149:© Dustin White_149
150:© Dustin White_150
151:© Dustin White_151
152:© Dustin White_152
153:© Dustin White_153
154:© Dustin White_154
155:© Dustin White_155
156:© Dustin White_156
157:© Dustin White_157
158:© Dustin White_158
159:© Dustin White_159
160:© Dustin White_160
161:© Dustin White_161
162:© Dustin White_162
163:© Dustin White_163
164:© Dustin White_164
165:© Dustin White_165
166:© Dustin White_166
167:© Dustin White_167
168:© Dustin White_168
169:© Dustin White_169
170:© Dustin White_170
171:© Dustin White_171
172:© Dustin White_172
173:© Dustin White_173
174:© Dustin White_174
175:© Dustin White_175
176:© Dustin White_176
177:© Dustin White_177
178:© Dustin White_178
179:© Dustin White_179
180:© Dustin White_180
181:© Dustin White_181
182:© Dustin White_182
183:© Dustin White_183
184:© Dustin White_184
185:© Dustin White_185
186:© Dustin White_186

187:© Dustin White_187
188:© Dustin White_188
189:© Dustin White_189
190:© Dustin White_190
191:© Dustin White_191
192:© Dustin White_192
193:© Dustin White_193
194:© Dustin White_194

Estrangements

Cesare Battelli

1:© Cesare Battelli_1
2:© Cesare Battelli_2
3:© Cesare Battelli_3
4:© Cesare Battelli_4
5:© Cesare Battelli_5
6:© Cesare Battelli_6
7:© Cesare Battelli_7
8:© Cesare Battelli_8
9:© Cesare Battelli_9
10:© Cesare Battelli_10
11:© Cesare Battelli_11
12:© Cesare Battelli_12
13:© Cesare Battelli_13
14:© Cesare Battelli_14
15:© Cesare Battelli_15
16:© Cesare Battelli_16
17:© Cesare Battelli_17
18:© Cesare Battelli_18
19:© Cesare Battelli_19
20:© Cesare Battelli_20
21:© Cesare Battelli_21
22:© Cesare Battelli_22
23:© Cesare Battelli_23
24:© Cesare Battelli_24
25:© Cesare Battelli_25
26:© Cesare Battelli_26
27:© Cesare Battelli_27
28:© Cesare Battelli_28
29:© Cesare Battelli_29
30:© Cesare Battelli_30
31:© Cesare Battelli_31
32:© Cesare Battelli_32
33:© Cesare Battelli_33
34:© Cesare Battelli_34
35:© Cesare Battelli_35
36:© Cesare Battelli_36
37:© Cesare Battelli_37
38:© Cesare Battelli_38
39:© Cesare Battelli_39

Virginia San Fratello

1:© Virginia San Fratello_1
2:© Virginia San Fratello_2
3:© Virginia San Fratello_3
4:© Virginia San Fratello_4
5:© Virginia San Fratello_5
6:© Virginia San Fratello_6
7:© Virginia San Fratello_7
8:© Virginia San Fratello_8
9:© Virginia San Fratello_9
10:© Virginia San Fratello_10
11:© Virginia San Fratello_11
12:© Virginia San Fratello_12
13:© Virginia San Fratello_13
14:© Virginia San Fratello_14

15:© Virginia San Fratello_15
16:© Virginia San Fratello_16
17:© Virginia San Fratello_17
18:© Virginia San Fratello_18
19:© Virginia San Fratello_19
20:© Virginia San Fratello_20
21:© Virginia San Fratello_21
22:© Virginia San Fratello_22
23:© Virginia San Fratello_23
24:© Virginia San Fratello_24
25:© Virginia San Fratello_25
26:© Virginia San Fratello_26
27:© Virginia San Fratello_27
28:© Virginia San Fratello_28
29:© Virginia San Fratello_29
30:© Virginia San Fratello_30
31:© Virginia San Fratello_31
32:© Virginia San Fratello_32
33:© Virginia San Fratello_33
34:© Virginia San Fratello_34
35:© Virginia San Fratello_35
36:© Virginia San Fratello_36
37:© Virginia San Fratello_37
38:© Virginia San Fratello_38
39:© Virginia San Fratello_39
40:© Virginia San Fratello_40
41:© Virginia San Fratello_41
42:© Virginia San Fratello_42
43:© Virginia San Fratello_43
44:© Virginia San Fratello_44
45:© Virginia San Fratello_45
46:© Virginia San Fratello_46
47:© Virginia San Fratello_47
48:© Virginia San Fratello_48
49:© Virginia San Fratello_49
50:© Virginia San Fratello_50
51:© Virginia San Fratello_51
52:© Virginia San Fratello_52
53:© Virginia San Fratello_53
54:© Virginia San Fratello_54
55:© Virginia San Fratello_55
56:© Virginia San Fratello_56
57:© Virginia San Fratello_57
58:© Virginia San Fratello_58
59:© Virginia San Fratello_59
60:© Virginia San Fratello_60
61:© Virginia San Fratello_61
62:© Virginia San Fratello_62
63:© Virginia San Fratello_63
64:© Virginia San Fratello_64
65:© Virginia San Fratello_65
66:© Virginia San Fratello_66
67:© Virginia San Fratello_67
68:© Virginia San Fratello_68
69:© Virginia San Fratello_69
70:© Virginia San Fratello_70
71:© Virginia San Fratello_71
72:© Virginia San Fratello_72
73:© Virginia San Fratello_73
74:© Virginia San Fratello_74
75:© Virginia San Fratello_75
76:© Virginia San Fratello_76
77:© Virginia San Fratello_77
78:© Virginia San Fratello_78
79:© Virginia San Fratello_79
80:© Virginia San Fratello_80
81:© Virginia San Fratello_81
82:© Virginia San Fratello_82
83:© Virginia San Fratello_83
84:© Virginia San Fratello_84
85:© Virginia San Fratello_85
86:© Virginia San Fratello_86

87:© Virginia San Fratello_87
88:© Virginia San Fratello_88
89:© Virginia San Fratello_89
90:© Virginia San Fratello_90
91:© Virginia San Fratello_91
92:© Virginia San Fratello_92
93:© Virginia San Fratello_90
91:© Virginia San Fratello_91
92:© Virginia San Fratello_92
93:© Virginia San Fratello_93
94:© Virginia San Fratello_94
95:© Virginia San Fratello_95
96:© Virginia San Fratello_96
97:© Virginia San Fratello_97
98:© Virginia San Fratello_98
99:© Virginia San Fratello_99
100:© Virginia San Fratello_100
101:© Virginia San Fratello_101
102:© Virginia San Fratello_102
103:© Virginia San Fratello_103
104:© Virginia San Fratello_104
105:© Virginia San Fratello_105
106:© Virginia San Fratello_106
107:© Virginia San Fratello_107
108:© Virginia San Fratello_108
109:© Virginia San Fratello_109
110:© Virginia San Fratello_110
111:© Virginia San Fratello_111
112:© Virginia San Fratello_112
113:© Virginia San Fratello_113
114:© Virginia San Fratello_114
115:© Virginia San Fratello_115
116:© Virginia San Fratello_116
117:© Virginia San Fratello_117
118:© Virginia San Fratello_118

Immanuel Koh

1:© Immanuel Koh, The Hands of An Architect, 2023.
2:© Immanuel Koh, The Hands of An Architect, 2023.
3:© Immanuel Koh, The Hands of An Architect, 2023.
4:© Immanuel Koh, The Hands of An Architect, 2023.
5:© Immanuel Koh, 3D Janus Panda, 2023.
6:© Immanuel Koh, 3D Janus Panda, 2023.
7:© Immanuel Koh, 3D Janus Panda, 2023.
8:© Immanuel Koh, 3D Janus Panda, 2023.
9:© Immanuel koh_9
10:© Immanuel koh_10
11:© Immanuel koh_11
12:© Immanuel koh_12
13:© Immanuel koh_13
14:© Immanuel koh_14
15:© Immanuel koh_15
16:© Immanuel koh_16
17:© Immanuel koh_17
18:© Immanuel koh_18
19:© Immanuel koh_19
20:© Immanuel koh_20
21:© Immanuel koh_21
22:© Immanuel koh_22
23:© Immanuel koh_23
24:© Immanuel koh_24
25:© Immanuel koh_25
26:© Immanuel koh_26
27:© Immanuel koh_27
28:© Immanuel koh_28
29:© Immanuel koh_29

30:© Immanuel koh_30
31:© Immanuel koh_31
32:© Immanuel koh_32
33:© Immanuel koh_33
34:© Immanuel koh_34
35:© Immanuel koh_35
36:© Immanuel koh_36
37:© Immanuel koh_37
38:© Immanuel koh_38
39:© Immanuel koh_39
40:© Immanuel koh_40
41:© Immanuel koh_41
42:© Immanuel koh_42
43:© Immanuel koh_43
44:© Immanuel koh_44
45:© Immanuel koh_45
46:© Immanuel koh_46
47:© Immanuel koh_47
48:© Immanuel koh_48
49:© Immanuel koh_49
50:© Immanuel koh_50
51:© Immanuel koh_51
52:© Immanuel koh_52
53:© Immanuel koh_53
54:© Immanuel koh_54
55:© Immanuel koh_55
56:© Immanuel koh_56
57:© Immanuel koh_57
58:© Immanuel koh_58
59:© Immanuel koh_59
60:© Immanuel koh_60
61:© Immanuel koh_61
62:© Immanuel koh_62
63:© Immanuel koh_63
64:© Immanuel koh_64
65:© Immanuel koh_65
66:© Immanuel koh_66
67:© Immanuel koh_67
68:© Immanuel koh_68
69:© Immanuel koh_69
70:© Immanuel koh_70
71:© Immanuel koh_71
72:© Immanuel koh_72
73:© Immanuel koh_73
74:© Immanuel koh_74
75:© Immanuel koh_75
76:© Immanuel koh_76
77:© Immanuel koh_77
78:© Immanuel koh_78
79:© Immanuel koh_79
80:© Immanuel koh_80
81:© Immanuel koh_81
82:© Immanuel koh_82
83:© Immanuel koh_83
84:© Immanuel koh_84
85:© Immanuel koh_85
86:© Immanuel koh_86
87:© Immanuel koh_87
88:© Immanuel koh_88
89:© Immanuel koh_89
90:© Immanuel koh_90
91:© Immanuel koh_91
92:© Immanuel koh_92
93:© Immanuel koh_93
94:© Immanuel koh_94
95:© Immanuel koh_95
96:© Immanuel koh_96
97:© Immanuel koh_97
98:© Immanuel koh_98
99:© Immanuel koh_99
100:© Immanuel koh_100
101:© Immanuel koh_101

102:© Immanuel koh_102
103:© Immanuel koh_103
104:© Immanuel koh_104
105:© Immanuel koh_105
106:© Immanuel koh_106
107:© Immanuel koh_107
108:© Immanuel koh_108
109:© Immanuel koh_109
110:© Immanuel koh_110
111:© Immanuel koh_111
112:© Immanuel koh_112
113:© Immanuel koh_113
114:© Immanuel koh_114
115:© Immanuel koh_115
116:© Immanuel koh_116
117:© Immanuel koh_117
118:© Immanuel koh_118
119:© Immanuel koh_119
120:© Immanuel koh_120
121:© Immanuel koh_121
122:© Immanuel koh_122
123:© Immanuel koh_123
124:© Immanuel koh_124
125:© Immanuel koh_125
126:© Immanuel koh_126
127:© Immanuel koh_127
128:© Immanuel koh_128
129:© Immanuel koh_129
130:© Immanuel koh_130
131:© Immanuel koh_131
132:© Immanuel koh_132
133:© Immanuel koh_133
134:© Immanuel koh_134
135:© Immanuel koh_135
136:© Immanuel koh_136
137:© Immanuel koh_137
138:© Immanuel koh_138
139:© Immanuel koh_139
140:© Immanuel koh_140
141:© Immanuel koh_141
142:© Immanuel koh_142
143:© Immanuel koh_143
144:© Immanuel koh_144
145:© Immanuel koh_145
146:© Immanuel koh_146
147:© Immanuel koh_147
148:© Immanuel koh_148
149:© Immanuel koh_149
150:© Immanuel koh_150
151:© Immanuel koh_151
152:© Immanuel koh_152
153:© Immanuel koh_153
154:© Immanuel koh_154
155:© Immanuel koh_155
156:© Immanuel koh_156
157:© Immanuel koh_157
158:© Immanuel koh_158
159:© Immanuel koh_159
160:© Immanuel koh_160
161:© Immanuel koh_161
162:© Immanuel koh_162
163:© Immanuel koh_163
164:© Immanuel koh_164
165:© Immanuel koh_165
166:© Immanuel koh_166
167:© Immanuel koh_167
168:© Immanuel koh_168
169:© Immanuel koh_169
170:© Immanuel koh_170
171:© Immanuel koh_171
172:© Immanuel koh_172
173:© Immanuel koh_173

174:© Immanuel koh_174
175:© Immanuel koh_175
176:© Immanuel koh_176
177:© Immanuel koh_177
178:© Immanuel koh_178
179:© Immanuel koh_179
180:© Immanuel koh_180
181:© Immanuel koh_181
182:© Immanuel koh_182
183:© Immanuel koh_183
184:© Immanuel koh_184
185:© Immanuel koh_185
186:© Immanuel koh_186
187:© Immanuel koh_187
188:© Immanuel koh_188
189:© Immanuel koh_189
190:© Immanuel koh_190
191:© Immanuel koh_191
192:© Immanuel koh_192

Elena Manferdini

1:© Elena Manferdini_1
2:© Elena Manferdini_2
3:© Elena Manferdini_3
4:© Elena Manferdini_4
5:© Elena Manferdini_5
6:© Elena Manferdini_6
7:© Elena Manferdini_7
8:© Elena Manferdini_8
9:© Elena Manferdini_9
10:© Elena Manferdini_10
11:© Elena Manferdini_11
12:© Elena Manferdini_12
13:© Elena Manferdini_13
14:© Elena Manferdini_14
15:© Elena Manferdini_15
16:© Elena Manferdini_16
17:© Elena Manferdini_17
18:© Elena Manferdini_18
19:© Elena Manferdini_19
20:© Elena Manferdini_20
21:© Elena Manferdini_21
22:© Elena Manferdini_22
23:© Elena Manferdini_23
24:© Elena Manferdini_24
24:© Elena Manferdini_24
25:© Elena Manferdini_25
26:© Elena Manferdini_26
27:© Elena Manferdini_27
28:© Elena Manferdini_28
29:© Elena Manferdini_29
30:© Elena Manferdini_30
31:© Elena Manferdini_31
32:© Elena Manferdini_32
33:© Elena Manferdini_33
34:© Elena Manferdini_34
35:© Elena Manferdini_35
36:© Elena Manferdini_36
37:© Elena Manferdini_37
38:© Elena Manferdini_38
39:© Elena Manferdini_39
40:© Elena Manferdini_40
41:© Elena Manferdini_41
42:© Elena Manferdini_42
43:© Elena Manferdini_43
44:© Elena Manferdini_44
45:© Elena Manferdini_45
46:© Elena Manferdini_46
47:© Elena Manferdini_47
48:© Elena Manferdini_48

49:© Elena Manferdini_49
50:© Elena Manferdini_50

Kyle Steinfeld

1:© Kyle Steinfeld_1
2:© Kyle Steinfeld_2
3:© Kyle Steinfeld_3
4:© Kyle Steinfeld_4
5:© Kyle Steinfeld_5
6:© Kyle Steinfeld_6
7:© Kyle Steinfeld_7
8:© Kyle Steinfeld_8
9:© Kyle Steinfeld_9
10:© Kyle Steinfeld_10
11:© Kyle Steinfeld_11
12:© Kyle Steinfeld_12
13:© Kyle Steinfeld_13
14:© Kyle Steinfeld_14
15:© Kyle Steinfeld_15
16:© Kyle Steinfeld_16
17:© Kyle Steinfeld_17
18:© Kyle Steinfeld_18
19:© Kyle Steinfeld_19
20:© Kyle Steinfeld_20
21:© Kyle Steinfeld_21
22:© Kyle Steinfeld_22
23:© Kyle Steinfeld_23
24:© Kyle Steinfeld_24
25:© Kyle Steinfeld_25
26:© Kyle Steinfeld_26
27:© Kyle Steinfeld_27
28:© Kyle Steinfeld_28
29:© Kyle Steinfeld_29
30:© Kyle Steinfeld_30
31:© Kyle Steinfeld_31
32:© Kyle Steinfeld_32
33:© Kyle Steinfeld_33
34:© Kyle Steinfeld_34
35:© Kyle Steinfeld_35
36:© Kyle Steinfeld_36
37:© Kyle Steinfeld_37
38:© An image generated with the promtpt “De Architectura, Marcus Vitruvius Pollio, 20 BC”